Spectral Theory of Automorphic Functions

Mathematics and Its Applications (*Soviet Series*)

Volume 51

Spectral Theory of Automorphic Functions

and Its Applications

by

ALEXEI B. VENKOV

Steklov Mathematical Institute, Leningrad, U.S.S.R.

KLUWER ACADEMIC PUBLISHERS

DORDRECHT / BOSTON / LONDON

Library of Congress Cataloging-in-Publication Data

Venkov, A. B.
[Spektral'nafa teorifa avtomorfnykh funktsiĭ. English]
Spectral theory of automorphic functions and its applications / Alexei B. Venkov.
p. cm. -- (Mathematics and its applications (Soviet series) ; 51)
Translation of: Spektral'nafa teorifa avtomorfnykh funktsiĭ.
ISBN 0-7923-0487-X
1. Spectral theory (Mathematics) 2. Functions, Automorphic.
I. Title. II. Series: Mathematics and its applications (Kluwer Academic Publishers). Soviet series ; 51.
QA322.V4613 1990
515'.9--dc20 89-20006

ISBN 0-7923-0487-X

Translated by N. B. Lebedinskaya

Published by Kluwer Academic Publishers,
P.O. Box 17, 3300 AA Dordrecht, The Netherlands.

Kluwer Academic Publishers incorporates
the publishing programmes of
D. Reidel, Martinus Nijhoff, Dr W. Junk and MTP Press.

Sold and distributed in the U.S.A. and Canada
by Kluwer Academic Publishers,
101 Philip Drive, Norwell, MA 02061, U.S.A.

In all other countries, sold and distributed
by Kluwer Academic Publishers Group,
P.O. Box 322, 3300 AH Dordrecht, The Netherlands.

Printed on acid-free paper

Printed in the Netherlands

The crux of events is in events themselves, and the words about them have meaning. The crux of Tao is not in Tao and the words about it are meaningless. If one knows that words are meaningless then each word will turn one to Tao. If one does not know that words are meaningless then even the most refined words will be all the same as an eyesore.

From 'Guan In-tzy'

SERIES EDITOR'S PREFACE

'Et moi, ..., si j'avait su comment en revenir, je n'y serais point allé.'

Jules Verne

The series is divergent; therefore we may be able to do something with it.

O. Heaviside

One service mathematics has rendered the human race. It has put common sense back where it belongs, on the topmost shelf next to the dusty canister labelled 'discarded non-sense'.

Eric T. Bell

Mathematics is a tool for thought. A highly necessary tool in a world where both feedback and non-linearities abound. Similarly, all kinds of parts of mathematics serve as tools for other parts and for other sciences.

Applying a simple rewriting rule to the quote on the right above one finds such statements as: 'One service topology has rendered mathematical physics ...'; 'One service logic has rendered computer science ...'; 'One service category theory has rendered mathematics ...'. All arguably true. And all statements obtainable this way form part of the raison d'être of this series.

This series, *Mathematics and Its Applications*, started in 1977. Now that over one hundred volumes have appeared it seems opportune to reexamine its scope. At the time I wrote

> "Growing specialization and diversification have brought a host of monographs and textbooks on increasingly specialized topics. However, the 'tree' of knowledge of mathematics and related fields does not grow only by putting forth new branches. It also happens, quite often in fact, that branches which were thought to be completely disparate are suddenly seen to be related. Further, the kind and level of sophistication of mathematics applied in various sciences has changed drastically in recent years: measure theory is used (non-trivially) in regional and theoretical economics; algebraic geometry interacts with physics; the Minkowsky lemma, coding theory and the structure of water meet one another in packing and covering theory; quantum fields, crystal defects and mathematical programming profit from homotopy theory; Lie algebras are relevant to filtering; and prediction and electrical engineering can use Stein spaces. And in addition to this there are such new emerging subdisciplines as 'experimental mathematics', 'CFD', 'completely integrable systems', 'chaos, synergetics and large-scale order', which are almost impossible to fit into the existing classification schemes. They draw upon widely different sections of mathematics."

By and large, all this still applies today. It is still true that at first sight mathematics seems rather fragmented and that to find, see, and exploit the deeper underlying interrelations more effort is needed and so are books that can help mathematicians and scientists do so. Accordingly MIA will continue to try to make such books available.

If anything, the description I gave in 1977 is now an understatement. To the examples of interaction areas one should add string theory where Riemann surfaces, algebraic geometry, modular functions, knots, quantum field theory, Kac-Moody algebras, monstrous moonshine (and more) all come together. And to the examples of things which can be usefully applied let me add the topic 'finite geometry'; a combination of words which sounds like it might not even exist, let alone be applicable. And yet it is being applied: to statistics via designs, to radar/sonar detection arrays (via finite projective planes), and to bus connections of VLSI chips (via difference sets). There seems to be no part of (so-called pure) mathematics that is not in immediate danger of being applied. And, accordingly, the applied mathematician needs to be aware of much more. Besides analysis and numerics, the traditional workhorses, he may need all kinds of combinatorics, algebra, probability, and so on.

In addition, the applied scientist needs to cope increasingly with the nonlinear world and the extra mathematical sophistication that this requires. For that is where the rewards are. Linear models are honest and a bit sad and depressing: proportional efforts and results. It is in the non-linear world that infinitesimal inputs may result in macroscopic outputs (or vice versa). To appreciate what I am hinting at: if electronics were linear we would have no fun with transistors and computers; we would have no TV; in fact you would not be reading these lines.

There is also no safety in ignoring such outlandish things as nonstandard analysis, superspace and anticommuting integration, p-adic and ultrametric space. All three have applications in both electrical engineering and physics. Once, complex numbers were equally outlandish, but they frequently proved the shortest path between 'real' results. Similarly, the first two topics named have already provided a number of 'wormhole' paths. There is no telling where all this is leading - fortunately.

Thus the original scope of the series, which for various (sound) reasons now comprises five subseries: white (Japan), yellow (China), red (USSR), blue (Eastern Europe), and green (everything else), still applies. It has been enlarged a bit to include books treating of the tools from one subdiscipline which are used in others. Thus the series still aims at books dealing with:

- a central concept which plays an important role in several different mathematical and/or scientific specialization areas;
- new applications of the results and ideas from one area of scientific endeavour into another;
- influences which the results, problems and concepts of one field of enquiry have, and have had, on the development of another.

Automorphic functions, i.e. 'nice' functions (in the analytic sense) which are invariant with respect to some discretely acting group, have been around for a quite long time. The subject has always been a meeting ground for various different areas in mathematics such as number theory, differential equations, harmonic analysis, algebraic groups, ... Since about 1956, a further branch of mathematics has become strongly involved: spectral theory; more precisely the spectral theory of automorphic Laplacians such as the Laplace Beltrami operator on the Lobachevsky plane. As the author writes there is now a recognizable area of mathematics which can be called Selberg theory though it is still very much in development and new applications to or relations with other parts of mathematics come up all the time (for instance, rather recently with string theory, quantum fields and quantum statistical mechanics).

I am therefore happy that this timely authoritative monograph on the present state of the spectral theory of automorphic functions is now appearing and I am most pleased that this takes place within the framework of this series.

The shortest path between two truths in the real domain passes through the complex domain.

J. Hadamard

La physique ne nous donne pas seulement l'occasion de résoudre des problèmes ... elle nous fait pressentir la solution.

H. Poincaré

Never lend books, for no one ever returns them; the only books I have in my library are books that other folk have lent me.

Anatole France

The function of an expert is not to be more right than other people, but to be wrong for more sophisticated reasons.

David Butler

Bussum, August 1990 Michiel Hazewinkel

Table of Contents

Preface

This book consists of three independent parts supplementing one another. The first main part is a large survey written especially for this edition. The second part (Appendix 1), 'Monodromy Groups and Automorphic Functions', is also an original survey. Finally, the third part (Appendix 2), 'Automorphic Functions for Effective Solutions of Certain Issues of the Riemann–Hilbert Problem', is a slightly modified paper published earlier in 'Zapisky nauchnykh seminarov LOMI', 162 (1987), pp. 5-42.

List of Notations

In asymptotic formulas we use the symbols O, o, and Ω in estimates for the order of the remaining terms; this notation is standard in analysis and in analytic number theory. We also use the the Vinogradov symbol $\ll$, which is analogous to O, but is sometimes more convenient. If the constants depend on some parameters, on ϵ, δ, say, we write $\underset{\epsilon,\delta}{\ll}$, $O_{\epsilon,\delta}$, or state it especially. The other notation includes $\mathbb{Z}$, $\mathbb{Q}$, $\mathbb{R}$, $\mathbb{C}$ for the ring of integers, the fields of rational, real and complex numbers respectively. For $z \in \mathbb{C}$, $|z|$ is the absolute value of z, $\arg z$ is the argument of z, $\arg z \in [0, 2\pi)$, $\operatorname{Re} z$, $\operatorname{Im} z$ are the real and imaginary parts of z, $\bar{z}$ is complex conjugate to z. In accordance with the traditions of one or another field of study we use different symbols in different sections to denote the same object, for example, $e(z) = \exp 2\pi i z = \mathrm{e}^{2\pi i z}$; we hope that this ambiguity does not lead to any misunderstanding.

$[a, b]$ $((a, b], [a, b), (a, b))$ is the set of all $x \in \mathbb{R}$ such that $a \le x \le b$ $(a < x \le b$, $a \le x < b$, $a < x < b$, respectively).

$$\mathbb{R}_+ = \text{the set of all } x \in \mathbb{R}, \quad x > 0.$$

$$\mathbb{Z}_+ = \text{the set of all } x \in \mathbb{Z}, \quad x > 0 \quad (\text{or } \mathbb{R}_+ \cap \mathbb{Z}).$$

$$[x] = \text{the greatest integer } \le x, \quad x \in \mathbb{R}.$$

$$\delta_{a,b} = \begin{cases} 1, & \text{if } a = b \quad \text{(Kronecker symbol)}, \\ 0, & \text{if } a \ne b. \end{cases}$$

$$\Gamma(s) = \text{the Euler gamma - function.}$$

$$\mathrm{li} = \text{the integral logarithm, } \mathrm{li}\, x = \int_2^x \frac{\mathrm{d}t}{t}, \quad x \in \mathbb{R}, \quad x \ge 2.$$

$$\mathbb{C}^n(\mathbb{R}^n) = \text{complex (real) linear space of dimension } n.$$

We use the notation which is standard in the U.S.S.R. for trigonometric, hyperbolic, and logarithmic functions: sin, cos, tg, ctg, sh, ch, th, cth, ln(log). $J_\nu(x)$, $H_\nu^{(1)}(x)$, $H_\nu^{(2)}$, $I_\nu(x)$, $K_\nu(x)$ are the various Bessel functions. $\mathrm{GL}(n, F) = \mathrm{GL}_n(F)$, $\mathrm{SL}(n, F) = \mathrm{SL}_n(F)$ are standard for the classical linear groups defined over the field F. $C_0^\infty(M)$, C^∞, $L_2(F; \mathrm{d}\mu)$ are standard for the functional spaces (see [7]).

For the map f in the space V, $\det f$ is the determinant, $\mathrm{Tr}\, f$ is the trace ($\mathrm{tr}\, f$, sometimes), $\ker f$ is the kernel, $\dim V$ is the dimension. For the integers m, N, k, $(m, N) = 1$ means that m, N are mutually prime; $m \equiv N (\mathrm{mod}\ k)$ is equivalent to $N - m$ is divisible by k. $\mathbb{A}_{\mathbb{F}}$ is the adèle ring of the field $\mathbb{F}$, $\mathbb{A}$ is the adèle ring of the field $\mathbb{Q}$. $\mathbf{e}(x) = \exp 2\pi i(x + \bar{x})$ (in Chapter 11 only), Res is the residue symbol for a meromorphic function.

CHAPTER 1

Introduction

The notion 'automorphic function' is a generalization of the notion 'periodic function'. It assumes the existence of a certain group acting discretely on the ordinary plane, for example, or on any other metric space. The function defined on this space, having sufficiently good analytic properties and being invariant relative to the action of the group, is called an automorphic function.

Although automorphic functions so defined have been known for many years, the term itself was proposed only in 1890 by Klein as an alternative to 'Fuchsian function', introduced by Poincaré. Both great mathematicians considered analytic automorphic functions for a special class of discrete groups of linear fractional transformations of a plane or a half-plane.

The simplest group of this kind is the modular group, i.e. the group of all second-order integer matrices with a unit determinant. This group, as well as certain modular functions, was also considered, possibly for the first time by Gauss, and it was Gauss who originated the classical period of the theory of automorphic functions. Then, during the nineteenth century and the first ten years of the twentieth century, automorphic functions appeared in different and completely unrelated fields of mathematics, as an aid to solving the most intriguing problems (inversing integrals, solving algebraic equations, studying number-theory functions, solving differential equations, the uniformization problem and so on). In this first period of development, the theory of automorphic functions culminated in the works of Klein and Poincaré.

At this time there was immense interest among mathematicians in automorphic functions and their applications. Fashion also probably played a part in this interest as the theory turned out to be perfectly in tune with the times. Automorphic function theory is a science for aesthetes (although it also appeals to pedants). The symmetry of geometric constructions, the plethora of elegant formulas, explicitly or exactly solvable problems and successful hypotheses* are very appealing aspects of the theory. (The creative work of the outstanding Dutch artist Escher, picturing in modern times (1930–1970) the likeable angels and devils filling out the various surfaces in a periodic (automorphic) and quasi-periodic way emphasizes the elegance of the theory.)

---

* See the survey by Fricke [81], covering the classical period of the theory of automorphic functions.

Then a rather sharp decrease in interest in automorphic functions took place that lasted until the 1950's. Mathematical interests changed. Questions such as justifying mathematics, further development of algebra, analytical number theory, general theory of differential equations and more abstract matters were put on the agenda. In this period, however, the development of automorphic function theory was successfully continued by the efforts of many outstanding mathematicians, the most important being Hecke, Siegel, Petersson and Maass.

A revival of the massive interest in automorphic functions occurred in contemporary times due to the work of Selberg (and, probably, because of the periodicity of fashion). This was the origin of the history of the spectral theory of automorphic functions, which is the subject of this book.

In his famous paper [164] (the first widely published paper on this topic), Atle Selberg introduced fundamental new ideas into the classical theory of automorphic forms. These ideas are connected with an extension of the earlier notion of an automorphic function or form. Instead of an analytic (meromorhpic) automorphic function which was studied by the classics, Selberg considered a mapping that is automorphic relative to a given finite-dimensional unitary representation of a discrete group and is an eigenfunction for a commutative ring of elliptic differential operators. At that time, Hans Maass' article [130] also appeared, containing similar non-analytic automorphic 'wave' functions defined in a special situation; however, it was Selberg who first took a serious look at Maass' work. In order to implement the new ideas, certain new techniques, not normally used in the classical theory of automorphic functions, were invoked. First, methods from the theory of selfadjoint operators in Hilbert space; then, methods from group representation theory over various fields, methods which turned out to be more natural in spaces of rank greater than one. It was the subsequent global development of Selberg's ideas in the setting of the representation theory of Lie groups which determined the true place of the classical theory of automorphic functions – at least in its number theoretic aspect – in the new, more general theory.

It is now possible to refer to the 'Selberg theory', or, in less personal terms, the spectral theory of automorphic functions, although this is still in its initial stages of development. It is very difficult to completely cover all the problems and to see the proper place of the results in the theory that is apparently epoch-making in the development of a considerable branch of mathematics. But we have attempted to carry out a preliminary classification of directions and results, as they exist at present. The foundations of the theory are:

(1) theorems on expansion in automorphic eigenfunctions of Laplacians defined on symmetric (weakly symmetric) Riemann spaces and studying their spectra;

(2) Eisenstein series theory;

(3) Selberg trace formulas; and

(4) the theory of the Selberg zeta-functions.

(Items (1)–(4) assume the preliminary development of the theory of reductive Lie groups, representations of them and of their discrete subgroups).

Here one should also indicate several very important applications of a theoretical nature of the results and methods from (1)–(4).

(5) applications to global problems in modern number theory, in particular, to studying the general reciprocity laws (the so-called 'Langlands philosophy');

(6) applications to solving some difficult problems in analytic number theory;

(7) applications to the theory of geometric and topological invariants of Riemannian manifolds; the spectral problem of moduli of Riemann surfaces;

(8) applications to classical (Dirichlet, Neumann) boundary value problems of mathematical physics in regular domains;

(9) Selberg theory and conformal (quasi-conformal) mappings of Riemann surfaces;

(10) applications to classical monodromy theory of ordinary differential equations;

(11) expansions in automorphic eigenfunctions of Laplacians as models of scattering theory;

(12) applications to quantum statistical mechanics, quantum field theory and to string theory.

At present, the list of applications of the Selberg theory is still incomplete. This theory also plays an important role in dynamic systems, in interpolation theory and, perhaps, in other fields unknown to us.

In this monograph we shall primarily be interested in questions concerning items (1)–(4), and parts of (5)–(8), we do not discuss practically items (8)–(12).

Our main viewpoint on the subject and the method of investigation is mathematical analysis in the sense of Whittaker and Watson.

The title of each Chapter is adequate to its contents. The main part of the monograph, namely Chapters 4–9, deals with two-dimensional spectral theory of automorphic functions, i.e. the Lobachevsky plane and Fuchsian groups of the first kind. Two-dimensional theory of automorphic functions inimitable in its characteristics was always standard since it is more clear, and hence contains more complete results, studied applications and connections with other fields of mathematics than the multi-dimensional theory.

Chapters 2 and 3 are introductory to the spectral theory of automorphic functions. Chapter 10 contains the solution of the concrete problem of classical number theory, ie. Kummer's problem on the distribution of Gauss cubic sums for prime arguments, by the methods of spectral theory of automorphic functions. It is the most technical chapter. Chapters 11 and 12 should be considered as a supplement to the theory developed in the rest of the book. We have presented here a survey of some general concepts and results of the basis of Selberg theory in multi-dimensional case and number-theoretic applications.

The main body of the text was written in 1984. During the last four years, progress has been made mainly in the shape of new ideas and results of the multi-dimensional Selberg theory. Some of these new results are considered in Chapter 13.

I would like to thank my colleagues Zograf and Proskurin, who are in fact my co-authors, for their help in writing this book.

CHAPTER 2

What Does One Need Automorphic Functions For? Some Remarks for a Pragmatic Reader

Our monograph is mainly devoted to the theory of nonanalytic automorphic functions* which are transcendental objects to the greater degree than the classical analytic automorphic functions or forms. However, it may be useful here to recall certain objective reasons that forced the classics to create the basis of automorphic functions. We selected here three classical theorems (more precisely, three fragments of theories), and in each of them automorphic functions play a central role, solving problems that have appeared in other fields of mathematics. For detailed proof and comments, we refer the reader to Chapter 13 where we give the necessary references.

Let us begin with applications to arithmetic. Automorphic functions are generating functions for certain, interesting, from an arithmetical point of view, numerical sequences and, thus, enable us to study these sequences. We clarify the assertion given here, but preliminarily recall some classical definitions. Let $H = \{z \in \mathbb{C}, \operatorname{Im} z > 0\}$ be the upper half-plane that is a symmetric Riemann homogeneous space $\mathrm{SL}(2,\mathbb{R})/\mathrm{SO}(2)$ in which the special discrete modular group $\Gamma_{\mathbb{Z}} = \mathrm{PSL}(2,\mathbb{Z})$ acts by isometric (linear-fractional) transformations (see Chapters 4, 8, list of notations). Let now $k \in \mathbb{Z}$. A function $f : H \to \mathbb{C}$ is called modular or automorphic, relative to the modular group, function (form) of weight $2k$, if it has the properties:

(1) f is meromorphic on H;
(2) for any

$$\gamma = \begin{pmatrix} a & b \\ c & d \end{pmatrix} \in \Gamma_{\mathbb{Z}}$$

and any $z \in H$ the equality

$$f(z) = (cz + d)^{-2k} f(\gamma z)$$

is valid;
(3) f is meromorphic at each point $q \in \mathbb{Q} \cup \{\infty\}$.

---

* See the general definition of automorphic function at the beginning of Chapters 1 and 4.

It is well-known that the modular group is generated by the following transformations $z \to -z^{-1}$, $z \to z+1$. Therefore condition (2) can be replaced by $f(z+1) = f(z)$, $f(-\frac{1}{z}) = f(z)z^{2k}$. This, in particular, implies that the modular function has a Fourier expansion

$$f(z) = \sum_{n=-l}^{\infty} a_n q^n = \sum_{n=-l}^{\infty} a_n \exp 2\pi i n z, \tag{2.1}$$

where $l \in \mathbb{Z}$. We shall call the modular form analytic if it is analytic in H and at each point $z = q \in \mathbb{Q} \cup \{\infty\}$ (by condition (2) it suffices for f to be analytic at ∞); it will be called a cusp-form or parabolic form, if f is an analytic modular function decreasing at infinity. In the expansion (2.1) we have $l = 0$ for an analytic modular function, and $l = -1$ for a cusp-form.

We introduce the notation

$$G_k(z) = \sum_{\substack{n,m \in \mathbb{Z} \\ (n,m) \neq (0,0)}} (nz+m)^{-2k}, \quad k \in \mathbb{Z}, \quad k \geq 1.$$

$G_k(z)$ is called an Eisenstein series. It may be shown that $G_k(z)$ is the analytic modular function of weight $2k$, and the expansion (2.1) has the form

$$G_k(z) = 2\varsigma(2k) + 2\frac{(2\pi i)^k}{(2k-1)!} \sum_{n=1}^{\infty} \sigma_{2k-1}(n) q^n, \tag{2.2}$$

where $\varsigma(s)$ is the Riemann zeta-function, $\sigma_a(n)$ is the sum of ath powers of positive divisors of the number n.

Note the following important algebraic assertion that any analytic modular function is a polynomial in two variables G_2, G_3 with complex coefficients.

There are no explicit formulas similar to (2.2) for Fourier coefficients of parabolic forms, but there are estimates for them with various, different degrees of precision. We only cite here the following classical Hecke theorem:

$$a_n \underset{n\to\infty}{=} O(n^k).$$

We now finally proceed to arithmetical applications in modular function theory as mentioned above. Consider the following sequence of numbers: for any $m \in \mathbb{Z}$, $m \geq 0$ we define $r_B(m)$ as the number of representations of $2m$ by the quadratic positive form $(Bx, x) = 2m$ in n variables ($n = 8l$). We define a generating function, the theta-series, as

$$\theta_B(z) = \sum_{m=0}^{\infty} r_B(m)\, e^{2\pi i m z},$$

with a view to study the coefficients $r_B(m)$ for large values. From Poisson's summation formula (see Chapter 3) it follows that $\theta_B(z)$ is the analytic modular function of weight $n/2$. The structural theorems of modular function theory (a brief account of which is given above) implies the theorem.

THEOREM 2.1. (1) *There exists a parabolic form f_B of weight $n/2$ such that*

$$\theta_B = \frac{1}{\varsigma(2k)} G_k + f_B.$$

(2) *The equality*

$$r_B(m) = \frac{4k}{B_k}\sigma_{2k-1}(m) + O(m^k), \quad k = n/4,$$

is valid, where B_k is the Bernulli number.

We now proceed to algebraic number theory. The question is Hilbert's twelfth problem in his famous report [104]: proving Kronecker's theorem on Abelian fields for the case of any algebraic rationality domain. Kronecker proved that any Abelian extension, i.e. a normal extension with a commutative Galois group, over the field of rational numbers $\mathbb{Q}$ is contained in the field generated by the roots of unit. Thus, a maximal Abelian extension over the field $\mathbb{Q}$ is generated by certain values of the special function $\exp(x)$. The question is whether for any algebraic number field k there exists a 'good' function which plays the role of $\exp(x)$ for describing the Abelian extensions over k. At present, the answer to Hilbert's question is not known in general case. However, the answer is well-known but not obvious if we take an imaginary quadratic field instead of $\mathbb{Q}$. In this case, a special modular function $j(z)$ plays the role of $\exp(x)$.

We define the function $j(z)$ (the normed modular invariant). Using the notation adopted above we set

$$g_2 = 60G_2, \quad g_3 = 140G_3, \quad \Delta = g_2^3 - 27g_3^2, \quad j = 1728g_2^3/\Delta.$$

The function $j(z)$ has the following analytical properties:

(1) j is modular of weight zero.

(2) j is holomorphic on H and has a simple pole at infinity with the residue equal to one.

(3) j realizes a bijection of the set $H/\Gamma_{\mathbb{Z}}$ onto $\mathbb{C}$.

In addition, $j(z)$ has certain important number-theoretic properties. Let k_i, $i = 1, 2, \dots, h$ be the ideal classes of a given imaginary quadratic field k.

THEOREM 2.2. (1) *The invariants of the classes $j(k_i)$ are algebraic integers.*

(2) *The field $k(j(k_i))$ does not depend on $i = 1, \dots, h$ and is a maximal Abelian extension over the field k.*

Any historical survey devoted to automorphic function theory will be incomplete if it does not contain the papers of Poincaré in analytical theory of ordinary differential equations. The great Poincaré, in particular, stated and solved the problem of explicit integration for ordinary linear differential equations with algebraic coefficients. Here we limit ourselves to the statement (not very strictly) of the following general assertion.

THEOREM 2.3. *With any ordinary linear differential equation with algebraic coefficients and simple singular points a group of the equation (the monodromy group) is connected, in terms of which (or in terms of its automorphic functions, if they exist)*

(*a*) *the equation can be integrated;*

(*b*) *one can obtain the total quality characteristic of its solutions.*

CHAPTER 3

Harmonic Analysis of Periodic Functions. The Hardy–Voronoĭ Formula

This short chapter can be regarded as an introduction to the problems to which the following chapters are devoted. In the comparatively simple case of a commutative discrete group acting on the Euclidean plane, we demonstrate some ideas and methods peculiar to the spectral theory of automorphic functions as a whole. We shall do this with one typical example. It is known that the famous Selberg trace formula, which plays a vital role in automorphic form theory, is historically preceded by its far simpler, 'commutative' analog, the so-called Hardy–Voronoĭ formula. Apparently, it was the circle problem that stimulated the derivation of the latter (we shall consider this well-known problem in analytical number theory later). It should be pointed out that the Hardy–Voronoĭ formula, which immediately follows from the Poisson summation formula, allows a totally different interpretation, much in the spirit of L. D. Faddeev's paper [34], from the point of view of spectral theory.

Although this approach does not lead to any new results, it is still useful, since it clarifies the ideas involved in the exposition which follows.

Finally, we return to the Hardy–Voronoĭ formula. Let $h(x)$ be an arbitrary rapidly decreasing function as $x \to \infty$, and let $r(n)$ be the number of representations of n as the sum of squares of two integers. The Hardy–Voronoĭ formula can be expressed as follows:

$$\sum_{n=0}^{\infty} r(n)h(n) = \pi \sum_{n=0}^{\infty} r(n) \int_0^{\infty} h(x) J_0(2\pi\sqrt{nx})\, dx, \tag{3.1}$$

where J_0 is the Bessel function.

We show how to derive (3.1) from the Poisson summation formula. Let $g : \mathbb{R}^2 \to \mathbb{R}$ be an infinite differentiable rapidly decreasing function and $\tilde{g}$ be its Fourier transform

$$\tilde{g}(a, b) = \int_{\mathbf{R}} \int_{\mathbf{R}} g(x, y) \overline{e(ax + by)}\, dx\, dy. \tag{3.2}$$

The Poisson summation formula has the form

$$\sum_{m,n \in \mathbf{Z}} g(m, n) = \sum_{m,n \in \mathbf{Z}} \tilde{g}(m, n). \tag{3.3}$$

We take $g(x,y) = h(x^2+y^2)$ with h from (3.1). Then the left-hand side of (3.3) coincides with the left-hand side of (3.1). If we change $x = \rho\cos\varphi$, $y = \rho\sin\varphi$, $\rho \in \mathbb{R}$, $0 \le \varphi < 2\pi$ in the integral (3.2), it is easy to see that the right-hand side of (3.3) coincides with the right-hand side of (3.1). This proves (3.1).

Now we provide the other proof of the Hardy–Voronoĭ formula. Let F be a torus considered as a factor-space $F = \Gamma\backslash\mathbb{R}^2$, where Γ is a rectangular lattice spanned on the vectors of the plane which are the complex numbers $1, i$. We introduce the standard Hilbert space $L_2(F)$ and consider a linear non-negative self-adjoint operator A in it generated by the differential operator Δ, where Δ is the Laplace operator; in rectangular coordinates

$$\Delta = \frac{\partial^2}{\partial x^2} + \frac{\partial^2}{\partial y^2}.$$

An expansion in eigenfunctions of the operator A is in fact a Fourier expansion. Then the resolvent $R(\lambda) = (A-\lambda)^{-1}$ of the operator A is given as an integral operator by its kernel $r(z,z';\lambda)$ $(z,z' \in F)$. For this kernel, the spectral expansion in the basis of eigenfunctions of the operator A is valid:

$$r(z,z';\lambda) = \sum_{k,l\in\mathbb{Z}} \frac{1}{4\pi^2(k^2+l^2)-\lambda}\overline{v_{kl}(z')}v_{kl}(z), \tag{3.4}$$

where $v_{kl}(z) = e(kx+ly)$, $x = \operatorname{Re} z$, $y = \operatorname{Im} z$, are eigenfunctions.

On the other hand, the kernel $r(z,z';\lambda)$ is expressed in the form

$$r(z,z';\lambda) = \sum_{\gamma\in\Gamma} G(z,\gamma z';s), \tag{3.5}$$

where $G(z,z';s)$ is Green's function of the equation $\Delta u + \lambda u = f$, $\lambda = s^2$, $\gamma z' = z' + m + in$, $\gamma = m + in$ for $m, n \in \mathbb{Z}$.

We clarify (3.5). Green's function $G(z,z';s)$ of the mentioned problem with the radiation condition is computed explicitly in terms of certain special functions (see [32])

$$G(z,z';s) = \tfrac{i}{4}H_0^2(|z-z'|s),$$

where H_0^2 is the Hankel function. The known asymptotic formula for the function $H_0^2(x)$ for a large value $|x|$ implies the absolute convergence of the series in the right-hand side of (3.5) in the region $\operatorname{Im} s < 0$. For s with $\operatorname{Im} s \ge 0$, the kernel $r(z,z';s)$ is the analytical continuation of this series. Now we let $h(\lambda)$ denote a function analytic in a neighbourhood of the ray $[0,\infty)$ and decreasing there as $O(|\operatorname{Re}\lambda|^{-k})$, $k \ge 2$. Consider an integral operator $h(A)$. We compute its spectral and matrix traces making use of (3.4), (3.5) and the known formula in functional analysis

$$h(A) = -\frac{1}{2\pi i}\int_\Omega h(\lambda)R(\lambda)\,d\lambda, \tag{3.6}$$

where Ω is a contour enclosing the spectrum and lying in the region of analyticity of the function h. One can prove that the operator $h(A)$ is of trace class with the mentioned choice of the function h. This enables us to equate the expressions obtained for the matrix and spectral traces. Deforming the Ω to a contour which consists of two rays (∞, io), $(-io, \infty)$, the equality

$$H_0^2(e^{\pi i}z) = H_0^2(z) + 2J_0(z)$$

and, finally, the obvious change in the variable lead to the desired formula (3.1).

This or the second proof lays restrictions on the function h which are considerably harder than those stated in (3.1). In addition, it is essentially more complicated than the first one. Nevertheless, this proof is very instructive, since it involves the ideas lying at the basis of deriving the Selberg trace formula.

We conclude this chapter by showing the connection of the Hardy–Voronoĭ formula with the circle problem. The problem is as follows. Let $D(X)$ be the number of solutions of the inequality $n^2 + m^2 \leq X$, $n, m \in \mathbb{Z}$. In other words, $D(X)$ is the number of points with integer coordinates in the circle of the radius $\sqrt{X}$. It is obvious that $X \geq 1$ and

$$D(X) = \pi X + R(X), \quad R(X) = O(\sqrt{X}).$$

It is necessary to find the smallest $\theta \in \mathbb{R}$ such that

$$R(X) = O(X^{\theta+\epsilon}) \tag{3.7}$$

for any $\epsilon > 0$. If we use the Hardy–Voronoĭ formula for studying $D(X)$ it is reasonable to take

$$h(x) = \max\{0, X - x\}.$$

This function is not differentiable at the point X, but after approximating it properly, one can see that (3.1) holds. From (3.1) we obtain the equality

$$\tilde{D}(X) = \tfrac{\pi}{2}X^{-2} + \tilde{R}(X)$$

with

$$\begin{aligned} \tilde{D}(X) &= \sum_{n\leq X}(X-n)r(n), \\ \tilde{R}(X) &= X/\pi \sum_{n=1}^{\infty} \frac{r(n)}{n} J_2(2\pi\sqrt{nX}). \end{aligned} \tag{3.8}$$

As

$$J_2(z) \ll z^{-1/2}, \quad z \to \infty,$$

and the series

$$\sum_{n=1}^{\infty} \frac{r(n)}{n^{5/4}}$$

converges, we have

$$\tilde{R}(X) \ll X^{3/4}.$$

To proceed to the estimate for $R(X)$ it suffices to note that

$$\tilde{D}(X) = \int_0^X D(t)\,\mathrm{d}t, \quad \tilde{R}(X) = \int_0^X R(t)\,\mathrm{d}t,$$

$$O(h) + 1/h \int_{X-h}^{X} R(t)\,\mathrm{d}t \le R(X) \le O(h) + 1/h \int_{X}^{X+h} R(t)\,\mathrm{d}t.$$

Our method gives (3.7) with $\theta = 3/8$. One can obtain a better result $\theta = 1/3$, if one makes use of asymptotics for $J_2(z)$ as $z \to \infty$, when estimating $\tilde{R}(X)$, and estimates the trigonometrical sums obtained.

Almost in such a way, one can obtain the best estimates for the value θ known at present (see Chapter 13). As for an expected value of θ, see also Chapter 13.

CHAPTER 4

Expansion in Eigenfunctions of the Automorphic Laplacian on the Lobachevsky Plane

We now proceed to exposing the basis of the spectral theory of automorphic functions, limiting ourselves for the present to the case of a hyperbolic plane (the Lobachevsky plane).

To avoid interrupting the subsequent exposition, we provide here some necessary facts from discrete group theory on the Lobachevsky plane.

We realize the Lobachevsky plane H as the upper half-plane $\{z = x + iy \in \mathbb{C} \mid y > 0\}$ in the complex plane $\mathbb{C}$ with the Poincaré metric (the Poincaré model)

$$ds^2 = \frac{1}{y^2}(dx^2 + dy^2). \tag{4.1}$$

The semicircles orthogonal to the real axis are the geodesics in this metric. The group of all motions without reflection of the upper half-plane H in the metric (4.1) coincides with the group $\mathrm{PSL}(2,\mathbb{R}) = \mathrm{SL}(2,\mathbb{R})/\{\pm 1\}$*. The element $\sigma \in \mathrm{SL}(2,\mathbb{R})$ acts on H by a linear-fractional transformation

$$z \to \sigma z = \frac{az+b}{cz+d}.$$

The matrices σ and $-\sigma$ are identified, since they define the same transformation of H.**

The elements of the group $\mathrm{PSL}(2,\mathbb{R})$ are elliptic, hyperbolic and parabolic, if the value of $|\mathrm{Tr}\,\sigma| = |a+d|$ is correspondingly less, more or equal to 2. We note that the elements of $\mathrm{PSL}(2,\mathbb{R})$ may also be considered as linear-fractional mappings of the extended complex plane $\overline{\mathbb{C}} = \mathbb{C} \cup \{\infty\}$ which transform the set $\mathbb{R} \cup \{\infty\}$ into itself. Any elliptic element has exactly two fixed points on $\mathbb{C}$ which are complex conjugate to each other, i.e. one fixed point on H. Hyperbolic elements also have two fixed points on $\overline{\mathbb{C}}$ belonging to the set $\mathbb{R} \cup \{\infty\}$. Parabolic elements have only one fixed point lying on $\mathbb{R} \cup \{\infty\}$.

---

* Note that H can be realized as a homogeneous space $\mathrm{SL}(2,\mathbb{R})/\mathrm{SO}(2)$ of the group $\mathrm{SL}(2,\mathbb{R})$ by its maximal compact subgroup $\mathrm{SO}(2)$.

** One does not usually distinguish between an element $\sigma \in \mathrm{PSL}(2,\mathbb{R})$ and its preimages in the group $\mathrm{SL}(2,\mathbb{R})$, since this does not lead to any misunderstanding.

A $PSL(2, \mathbb{R})$–invariant measure on H is connected with the Poincaré metric (4.1) and is described by the formula

$$d\mu(z) = \frac{1}{y^2}\, dx\, dy. \tag{4.2}$$

An invariant measure of a measurable subset F of the upper half-plane H is defined by $|F| = \int_F d\mu$. Later on, under a measure of a set we shall refer namely to an invariant measure, unless otherwise stated.

Under a discrete group of motions of the upper half-plane H we understand a subgroup Γ in $PSL(2, \mathbb{R})$ discrete in the sense of the induced topology. A point $z \in H$ is called an elliptic point of a discrete group Γ if there exists an element $\sigma \in \Gamma$, $\sigma \neq 1$ such that $\sigma z = z$; the group $\Gamma_z = \{\sigma \in \Gamma \mid \sigma z = z\}$ is a cyclic subgroup in Γ of finite order and is generated by a certain elliptic element. A point $x \in \mathbb{R} \cup \{\infty\}$ is called a parabolic point of the group Γ, if $\sigma x = x$ for a certain parabolic element $\sigma \in \Gamma$; the subgroup $\Gamma_x = \{\sigma \in \Gamma \mid \sigma x = x\}$ is in this case an infinite cyclic subgroup in Γ.

We now introduce a notion of a fundamental domain of a discrete group. A domain F on the upper half-plane H is called a fundamental domain of a discrete group Γ, if

(1) no two points of F are equivalent relative to Γ (i.e., there is no element in Γ which moves them into one another);

(2) any point on H is equivalent relative to Γ to a certain point of the closure of the domain F.

The fundamental domain exists for any discrete group acting on the Lobachevsky plane. We shall consider only those discrete groups, where the fundamental domains have a finite invariant measure. For these the spectral theory of automorphic functions is the most interesting. As Siegel showed [172], the class of such groups coincides exactly with the class of Fuchsian groups of the first kind. In other words, one can give the following simple definition: a Fuchsian group of the first kind is a discrete group of motions of the Lobachevsky plane for which there exists a fundamental domain with a finite invariant measure (see Chapter 13).

We note some particular properties of Fuchsian groups of the first kind that single them out from other discrete groups. According to the classical result of Fricke, any Fuchsian group Γ of the first kind is finitely generated. Furthermore, it may be given by a certain standard system of generators $A_1, B_1, \ldots, A_g, B_g, R_1, \ldots, R_l$, $S_1, \ldots, S_h \in PSL(2, \mathbb{R})$ and relations

$$A_1^{-1}B_1^{-1}A_1B_1 \cdots A_g^{-1}B_g^{-1}A_gB_gR_1 \cdots R_lS_1 \cdots S_h = 1,$$
$$R_j^{m_j} = 1, \quad j = 1, \ldots l \quad (m_j \in \mathbb{Z}, m_j \geq 2);$$

here the elements $A_1, B_1, \ldots, A_g, B_g$ are hyperbolic, $R_1, \ldots, R_l$ are elliptic, $S_1, \ldots, S_h$ are parabolic. The set of numbers $(g; m_1, \ldots, m_l; h)$ is an invariant of the group Γ and it is called its signature.

For any standard system of generators, a fundamental domain of a special kind can be chosen in the form of a convex (in the sense of non-Euclidean geometry) polygon bounded by $4g+2l+2h$ geodesics, the sides of which are pairwise identified under the action of the generators of the group. Such a fundamental domain will be called a normal polygon connected with the Fuchsian group Γ of the first kind. We note that the measure of the fundamental domain can be computed in terms of the signature by the Gauss–Bonnet formula

$$|F| = 2\pi\Big(2g-2+\sum_{j=1}^{l}(1-\tfrac{1}{m_j})+h\Big). \tag{4.3}$$

If $h=0$, then the fundamental domain of the group Γ is relatively compact in H; i.e., it has a compact closure; such a group Γ is called co-compact. If, in addition, we have $l=0$ then the group Γ is called strictly hyperbolic, and in this case the factor space $\Gamma\backslash H$ of the upper half-plane by the action of the group Γ is a compact Riemann surface of genus g. In the case when $h\neq 0$, a normal polygon F connected with the group Γ has exactly h vertices lying on $R\cup\{\infty\}$. These vertices are parabolic points of the group Γ and are pairwise inequivalent. We denote by x_k the vertex of F that is stabilized by a parabolic generator S_k of the group Γ. The polygon F can be represented as the union $F_0\cup\big(\cup_{k=1}^{h}F_k\big)$ of disjoint sets, where F_0 is a relatively compact region with a piecewise smooth boundary, and each F_k is the image of the strip $\Pi_Y=\{z\in H \mid |x|<1/2, y>Y\}$ with a sufficiently large $Y>0$ under the suitable linear-fractional mapping $\sigma_k\in \mathrm{PSL}(2,\mathbb{R})$ which transfers the point ∞ into x_k. The subgroup Γ_k in Γ stabilizing the point x_k is generated by S_k and, in addition,

$$\sigma_k^{-1}\Gamma_k\sigma_k=\left\{\begin{pmatrix}1 & n\\ 0 & 1\end{pmatrix}\,\middle|\, n\in\mathbb{Z}\right\}.$$

We conclude this short excursus to the theory of Fuchsian groups and proceed to the spectral theory.

First of all, we clarify some of the spectral properties of the Laplace (Laplace–Beltrami) operator on the Lobachevsky plane. The Laplace operator L associated with the Poincaré metric (4.1) on the upper half-plane H and taken with the sign minus is expressed explicitly in the form

$$L\equiv -y^2\left(\frac{\partial^2}{\partial x^2}+\frac{\partial^2}{\partial y^2}\right).$$

This operator is invariant, i.e. commutative with all operators of the quasi-regular representation of the group $\mathrm{PSL}(2,\mathbb{R})$ in the space $C_0^\infty(H)$.

We consider the algebra of all invariant operators (integral and differential of finite order) acting in the space mentioned above. As Selberg showed in his paper [164], this algebra is commutative and, in addition, any of its elements are a function of the operator L. Some of the geometric properties of the Lobachevsky plane lie at the base of mentioned assertions, and, especially, the fact that H is a symmetric space of rank 1. We shall see later just how significant these assertions are.

Consider an arbitrary integral operator K which is given by a kernel $k(z,z')$. Invariance of the operator is equivalent to fulfilling the condition $k(\sigma z, \sigma z') = k(z,z')$ for any $z, z' \in H$ and $\sigma \in \mathrm{PSL}(2,\mathbb{R})$. In other words, the kernel of the invariant operator is a function which depends only on the geodesic distance $d(z,z')$ between the points z, z'. To avoid technical difficulties, it is more convenient to replace the distance $d(z,z')$ by the so-called 'fundamental invariant of a pair of points' defined as follows:

$$t(z,z') = 2(\operatorname{ch} d(z,z') - 1) = |z - z'|^2/yy',$$

and to assume that $k(z,z') = k(t(z,z'))$. As was noted above, there exists a function $\widetilde{h}$ such that the equality $K = \widetilde{h}(L)$ holds. Leaving apart the analytical properties of the functions k and $\widetilde{h}$, it is natural to clarify the formulas that connect them. Such formulas were discovered by Selberg [164]:

$$\begin{gathered} \int_w^\infty \frac{k(t)}{\sqrt{t-w}}\,\mathrm{d}t = Q(w), \quad k(t) = -1/\pi \int_t^\infty \frac{\mathrm{d}Q(w)}{\sqrt{w-t}}, \\ Q(\mathrm{e}^u + \mathrm{e}^{-u} - 2) = g(u) \\ h(r) = \int_{-\infty}^{+\infty} \mathrm{e}^{iru} g(u)\,\mathrm{d}u, \quad g(u) = 1/2\pi \int_{-\infty}^{\infty} \mathrm{e}^{-iru} h(r)\,\mathrm{d}r, \end{gathered} \tag{4.4}$$

where $h(r) = \widetilde{h}(1/4 + r^2)$. It is clear that these relations hold true provided that $k \in C_0^\infty([0,\infty))$, for example. Formula (4.4) is, in essence, a special case of a theorem on expansion of an arbitrary spherical function in zonal functions (the Fourier–Harish–Chandra transform) which is valid in considerably more general cases than the upper half-plane. Relations (4.4) play a significant role in deriving the Selberg trace formula. We note that it is not difficult to obtain them if we consider the values of the operator K on a sufficiently representative set of functions, for example, on the set of functions y^s, $s \in \mathbb{C}$.

We now give several definitions of automorphic spectral theory. Let Γ be a Fuchsian group of the first kind, which we consider as a group of linear-fractional transformations of the upper half-plane H, and χ be a unitary representation of the group Γ in a Hermitian space V, $\dim V = n$. A vector-valued function $f : H \to V$ is called automorphic relative to Γ and χ (or, simply, automorphic when it is clear which Γ and χ the question is about) if for any $\gamma \in \Gamma$ and $z \in H$ the condition

$$f(\gamma_z) = \chi(\gamma) f(z)$$

holds.

We now fix a sufficiently good fundamental domain F of the group Γ (a normal polygon, for example) and consider a standard Hilbert space $L_2(F; V; \mathrm{d}\mu)$ of vector-functions with values in V square integrable on F relative to the measure $\mathrm{d}\mu$. The inner product in this space is given by the formula

$$(f_1, f_2) = \int_F \langle f_1, f_2\rangle_V \, d\mu,$$

where $\langle\,,\rangle_V$ is an Hermitian form in V. For convenience, we let $\mathcal{H} = \mathcal{H}(\Gamma;\chi)$ denote the space $L_2(F;V;d\mu)$.

To define an automorphic Laplacian $A(\Gamma;\chi)$ for a group Γ and a representation χ, we first consider the Laplace operator L on the set D of all smooth automorphic functions f given on H such that the restriction of f on the fundamental domain lies in the space $\mathcal{H}(\Gamma;\chi)$ along with Lf. It is evident that the set D is dense in $\mathcal{H}(\Gamma;\chi)$, and the operator L is symmetric and non-negative on this set. It is well known that in this case the operator L admits a self-adjoint extension on the space $\mathcal{H}$ (the so-called Friedrichs extension).

Anticipating what comes later, we note that the self-adjoint extension of the operator L with the domain of definition D in the space $\mathcal{H}$ is, in fact, unique (this follows from the properties of its resolvent).

We shall call a so-called self-adjoint non-negative unbounded linear operator $A = A(\Gamma;\chi)$ in the space $\mathcal{H} = \mathcal{H}(\Gamma;\chi)$ an automorphic Laplacian.

We describe some of the spectral properties of the operator $A(\Gamma;\chi)$. We first single out one simple, special case when the group Γ is co-compact.

In this case $A(\Gamma;\chi)$ is the operator with a purely discrete spectrum. In other words, the spectrum of the operator $A(\Gamma;\chi)$ consists only of some isolated eigenvalues of finite multiplicity, and in the space $\mathcal{H}$ it is possible to choose an eigenbasis. This fact can be easily proved using some general methods of functional analysis (more precisely, making use of a proper embedding theorem).

We now proceed to the spectral expansion of an invariant integral operator in the space of automorphic functions. We associate with any real-valued function $k \in C_0^\infty([0,\infty))$ an integral operator $K_{\Gamma,\chi}$ in the space $\mathcal{H}(\Gamma;\chi)$, the kernel of which is defined by the formula

$$k_{\Gamma,\chi}(z,z') = \sum_{\gamma\in\Gamma} \chi(\gamma) k(t(z,\gamma z')), \quad z,z' \in H. \tag{4.5}$$

If f is an automorphic function on H, then, as can be seen,

$$(K_{\Gamma,\chi} f)(z) = \int k_{\Gamma,\chi}(z,z') f(z')\, d\mu(z') = \int_H k(z,z') f(z')\, d\mu(z').$$

We denote by $0 \le \lambda_0 \le \lambda_1 \le \cdots$ the ordered set of eigenvalues of the operator $A(\Gamma;\chi)$, taken with their multiplicity, and by $\{w_j\}_{j=0}^\infty$ the corresponding orthonormal eigenbasis of the space $\mathcal{H}(\Gamma;\chi)$ consisting of eigenfunctions of the operator $A(\Gamma;\chi)$. The following spectral expansion is valid (see [161], [151]):

$$k_{\Gamma,\chi}(z,z') = \sum_j \widetilde{h}(\lambda_j) w_j(z) \otimes \overline{w_j(z')}, \tag{4.6}$$

where the function $\widetilde{h}$ is related to the kernel k by means of (4.4), $\otimes$ is the tensor product of values of vector-functions (written in the form of an $n \times n$ square matrix). This expansion means, in essence, the validity of the following operator equality

$$K_{\Gamma,\chi} = \widetilde{h}(A(\Gamma;\chi)).$$

The proof of (4.6) can be obtained in various ways, for example, using the general formula (3.6) for the resolvent of the automorphic Laplacian (see [34]).

We now suppose that Γ is a Fuchsian group of the first kind with a non-compact fundamental domain. As will be seen later, the spectral theory becomes much more complex in this case.

By analogy with the beginning of this chapter, we denote by $x_1, \dots x_h$ the set of pairwise inequivalent parabolic vertices of the fundamental domain F of Γ (i.e. the set of inequivalent vertices of F lying on $\mathbb{R} \cup \{\infty\}$), and by S_k the generator of the subgroup $\Gamma_k \in \Gamma$ which stabilizes the vertex x_k $(k = 1, \dots, h)$.

A finite-dimensional unitary representation χ of the group Γ is singular in the vertex x_k if $\dim \ker(\chi(S_k) - 1_V) \neq 0$, where 1_V is the identity operator in the space of the representation χ. If the representation χ is singular in at least one of the cusps of the fundamental domain F, then it is called singular; it is non-singular (or regular*) otherwise.

In the cases where χ is non-singular, the operator $A = A(\Gamma;\chi)$ defined above has a purely discrete spectrum. This assertion is due to Selberg (see [164]). This can easily be proved by studying asymptotic behaviour of the kernel of the resolvent for the operator A in neighbourhoods of cusps of Γ. The desired assertion about the spectrum of A follows from compactness of the resolvent as an integral operator. Thus, the spectral theory of the operator $A(\Gamma;\chi)$ which corresponds to a non-singular representation χ differs little from that which we have presented above for a co-compact Fuchsian group Γ.

Let χ now be a singular representation. In this case, the operator $A = A(\Gamma;\chi)$ has a continuous spectrum, and in the space $\mathcal{H} = \mathcal{H}(\Gamma;\chi)$ one can no longer choose an eigenbasis for the operator A. How can we construct the spectral expansion of A in this case? This problem was solved by Selberg. He was successful in finding a subspace of $\mathcal{H}$, in which the operator A has a purely discrete spectrum. In the orthogonal complement to this subspace, it appears to be possible to describe the operator A in terms of the so-called Eisenstein–Maass series – we shall discuss this in detail below. More precisely, Selberg showed that in the case considered here, the spectrum of the operator A consists of a discrete part and a part which is absolutely continuous and of a finite multiplicity, and described the eigenfunctions of the continuous spectrum of the operator A (see Chapter 13).

Concerning the remaining content of this chapter, we will be giving, for the first time, an outline of the proof of the spectral theorem for the operator A in the spirit

---

* Do not confuse with a regular representation of a Lie group (see Chapter 11).

of Selberg's ideas.

We will then comment briefly on the contents of the paper by Faddeev [34] in which he proves the theorem on expansion of the automorphic Laplacian in eigenfunctions using the theory of perturbations of the continuous spectrum, well-known in modern mathematical physics. The main idea of [34] is as follows: the information needed for constructing the spectral expansion of the operator A (including the description of eigenfunctions of its continuous spectrum) is contained entirely in its resolvent $R(\lambda) = (A-\lambda)^{-1}$. Such an approach is a useful addition to the classical proof of Selberg and clarifies many of his ideas (see Chapter 13).

Thus, let Γ be a Fuchsian group of the first kind with a non-compact fundamental domain, χ be its unitary singular finite-dimensional representation. We are interested in spectral expansions of the operators $K_{\Gamma,\chi}$ given by the kernels of the form (4.5).

To avoid overloading the remainder of the exposition, we suppose that for the Fuchsian group Γ we have chosen its fundamental domain F has only one cusp, the point ∞, and the representation χ is one-dimensional, trivial ($\chi \equiv 1$), and thus, singular. For more general Γ and χ the corresponding results will be stated in the final form at the end of the chapter.

One may assume, not diminishing the generality, that the fundamental domain F of the group Γ on the upper half-plane H is bounded by geodesics $x=0$ and $x=1$. The cusp of a fundamental domain, such as that, is stabilized by the subgroup $\Gamma_\infty \subset \Gamma$ which is generated by the transformation $S: z \to z+1$, $z \in H$.

For a function ψ in the space $C_0^\infty([0,\infty))$, we define an automorphic function θ_ψ (a so-called incomplete theta-series) by the formula

$$\theta_\psi(z) = \sum_{\gamma\in\Gamma_\infty\backslash\Gamma} \psi(y(\gamma_z)),$$

where $y(z) = \operatorname{Im} z$, $z \in H$. It is evident that $\theta_\psi \in \mathcal{H}$. The closure of the subspace of all incomplete theta-series in $\mathcal{H}$ is denoted by Θ.

Let f be an arbitrary element in $\mathcal{H}$. We consider in $\mathcal{H}$ an inner product

$$(\theta_\psi, f) = \int_F \theta_\psi(z)\overline{f(z)}\,\mathrm{d}\mu(z) = \int_{\Gamma_\infty\backslash H} \psi(y)\overline{f(z)}\,\mathrm{d}\mu(z)$$
$$= \int_0^\infty \psi(y) \int_0^1 \overline{f(z)}\,\mathrm{d}x \frac{\mathrm{d}y}{y^2}.$$

The equality

$$\int_0^1 f(z)\,\mathrm{d}x = 0$$

for all $y > 0$ (i.e. identical vanishing the constant term in the Fourier-series expansion for the function f by the subgroup Γ_∞*) is a necessary and sufficient condition

* For Fourier expansions of automorphic functions, see Chapter 5.

for orthogonality of f to the subspace Θ.

Those functions f are usually called parabolic forms of weight zero (or cusp-functions). We let $\mathcal{H}_0$ denote the closed subspace in $\mathcal{H}$ of all parabolic forms. Evidently, $\mathcal{H} = \Theta \oplus \mathcal{H}_0$. It is easy to show that the projection operators in $\mathcal{H}$ onto the subspaces Θ and $\mathcal{H}_0$ commute with the operators $K_{\Gamma,\chi}$ and, so, with the operator A. We note that the operators $P_0 K_{\Gamma,\chi}$ where P_0 is the orthogonal projection in $\mathcal{H}$ onto the subspace $\mathcal{H}_0$, are compact (we assume as before that the kernel of the operator $K_{\Gamma,\chi}$ is given by means of a finite function k making use of (4.5)). Hence it follows immediately that the operator A has a purely discrete spectrum in the subspace $\mathcal{H}_0$.

In order to describe the spectral expansion in the subspace Θ, we consider an Eisenstein–Maass series

$$E(z,s) = \sum_{\sigma \in \Gamma_\infty \backslash \Gamma} y^s(\sigma z), \quad y(z) = \operatorname{Im} z.$$

This series converges absolutely for all of the values $s \in \mathbb{C}$, such that $\operatorname{Re} s > 1$ and uniformly in z on compact subsets of the upper half-plane H. Its sum $E(z,s)$ for each s as this is a real analytic function on H automorphic relative to the group Γ and satisfies the differential equation $LE(z,s) = s(1-s)E(z,s)$. $E(z,s)$ is holomorphic in s in the region $\operatorname{Re} s > 1$. One can show (and this is the most serious aspect of spectral theory that we are discussing) that $E(z,s)$ admits meromorphic continuation onto the entire complex s-plane. There are several proofs for this assertion; some corresponding remarks are in Chapter 13. We will now comment on the possible proof originated by Selberg [165] and described in detail in Kubota [117].

Consider the constant term of the Fourier expansion of the Eisenstein series $E(z,s)$ by the group Γ_∞ which, as the calculations show, is equal to

$$y^s + \varphi(s) y^{1-s}.$$

The function $\varphi(s)$ is given on the half-plane $\operatorname{Re} s > 1$ by a generalized Dirichlet series (see Chapter 5)

$$\varphi(s) = \pi^{1/2} \frac{\Gamma(s-\frac{1}{2})}{\Gamma(s)} \sum_{c>0} \sum_{0 \le d < c} c^{-2s},$$

where $\Gamma(s)$ is the Euler gamma-function, and summation is taken over all pairs c,d such that $\begin{pmatrix} * & * \\ c & d \end{pmatrix} \in \Gamma_\infty \backslash \Gamma / \Gamma_\infty$ and $c > 0$. The function $\varphi(s)$ plays an exclusive significant role in Selberg's theory as a whole, and soon this will have to be proved. For the present, we define a modified Eisenstein series $E^Y(z,s)$ on the fundamental domain F by the formula

$$E^Y(z,s) = \begin{cases} E(z,s), & y \le Y, \\ E(z,s) - y^s - \varphi(s) y^{1-s}, & y > Y. \end{cases}$$

The function $E^Y(z,s)$ so defined belongs to the space $\mathcal{H}$ for all values s in the region $\operatorname{Re} s > 1$ ($E(z,s) = O(y^{\max\{\operatorname{Re} s,\, 1-\operatorname{Re} s\}})$ as will be seen from the results of Chapter 5. It is not difficult to obtain the validity of the following relation

$$(E^Y(z,s), \overline{E^Y(z,s')}) = \frac{Y^{s+s'-1} - \varphi(s)\varphi(s')Y^{-s-s'+1}}{s+s'-1} + \\ + \frac{\varphi(s')Y^{s-s'} - \varphi(s)Y^{-s+s'}}{s-s'} \tag{4.7}$$

by means of Green's formula (or the formula of integrating by parts).

Formula (4.7) is the Maass–Selberg relation.

We now outline the idea of proving meromorphy of functions $\varphi(s)$ and $E(z,s)$. First, certain general arguments enable us to prove the existence of meromorphic continuation of mentioned functions to the region $\operatorname{Re} s > \frac{1}{2}$, $s \notin (\frac{1}{2}, 1]$. In terms of the variable $\lambda = s(1-s)$ this region coincides exactly with the λ-plane with the semi-axis $[0,\infty)$ removed, on which the spectrum of the operator A lies. Here we use the fact that the resolvent of any self-adjoint operator is an analytic function everywhere outside the spectrum of this operator. We shall not dwell on proving this fact. However, we note that the Maass–Selberg relation makes sense for all values of s in the region mentioned above. This relation plays a vital role in further proof. Namely, it enables us to obtain the existence of finite limits of the functions $\varphi(s)$ and $E(z,s)$ as $\operatorname{Re} s$ approaches $\frac{1}{2}$. Furthermore, by means of (4.7), one can show that $\varphi(s)$ and $E(z,s)$ may have only a finit number of simple poles in the interval $(\frac{1}{2}, 1]$, and their poles always coincide (note that the poles of $E(z,s)$ do not depend on z). From the Maass–Selberg relation, we also get a functional equation for the function $\varphi(s)$. It follows from the necessary condition for unimodularity of the function φ on the line $\operatorname{Re} s = \frac{1}{2}$:

$$\varphi(\tfrac{1}{2} + ir)\varphi(\tfrac{1}{2} - ir) = |\varphi(\tfrac{1}{2} + ir)|^2 = 1.$$

The symmetry principle enables us to extend the function $\varphi(s)$ to the entire complex plane, where it satisfies the functional equation

$$\varphi(s)\varphi(1-s) = 1. \tag{4.8}$$

In just the same way, one can continue the function $E(z,s)$ to the entire s-plane, and, in addition,

$$E(z,s) = \varphi(s)E(z,1-s). \tag{4.9}$$

Now we are able to describe the spectral expansion of the operator A in the subspace Θ. We pick out in Θ a finite-dimensional subspace Θ_0 spanned by the residues of the Eisenstein series $E(z,s)$ at those poles which lie in the interval $(\frac{1}{2}, 1]$. We denote by Θ_1 the orthogonal complement in Θ to the subspace Θ_0. For each function θ_ψ defined above which belongs to the subspace Θ_1 the following remarkable formula

$$(K_{\Gamma,\chi}\theta_\psi)(z) = \frac{1}{4\pi i}\int_{\mathrm{Re}\, s=\frac{1}{2}} h(s)(\theta_\psi, E(\cdot, s))E(z,s)\, \mathrm{d}s \tag{4.10}$$

holds. This follows from the definition of the function $E(z,s)$, its meromorphy and the classical Cauchy theorem in complex analysis.

We note that the inner product in the right-hand side of this formula makes sense although the function $E(z,s)$ does not belong to the space $\mathcal{H}$. It is not hard to see from (4.10) that each of the operators $K_{\Gamma,\chi}$ and, consequently, the operator A has in the subspace Θ_1 a purely continuous spectrum.

In summary, we can say that in the case of the Fuchsian group with one cusp and its trivial one-dimensional representation the space $\mathcal{H}$ can be decomposed into a direct sum of the subspaces $\mathcal{H}_0$, Θ_0, Θ_1, invariant for the operator A. The spectrum of the operator A is purely discrete in the subspace $\mathcal{H}_0 \oplus \Theta_0$ and absolutely continuous in the subspace Θ_1, filling up the semi-axis $[\frac{1}{4}, \infty)$ one time. A restriction to the line $\mathrm{Re}\, s = \frac{1}{2}$ of the meromorphically continued Eisenstein series $E(z,s)$ is in this case an eigenfunction of the continuous spectrum of the operator A.

We are able to expand an arbitrary function in eigenfunctions of the automorphic Laplacian in the following way

$$f(z) = \frac{1}{4\pi}\int_{-\infty}^{\infty} a(r)E(z, \tfrac{1}{2} + ir)\, \mathrm{d}r + \sum_j b_j w_j(z),$$

where $\{w_j\}$ is an orthonormal eigenbasis in $\mathcal{H}_0 \oplus \Theta_0$,

$$a(r) = \int_F f(z)\overline{E(z, \tfrac{1}{2} + ir)}\, \mathrm{d}\mu(z), \quad b_j = \int_F f(z)\overline{w_j(z)}\, \mathrm{d}\mu(z);$$

in addition, the Parseval equality

$$\int_F |f(z)|^2\, \mathrm{d}\mu(z) = \frac{1}{4\pi}\int_{-\infty}^{\infty} |a(r)|^2\, \mathrm{d}r + \sum_j |b_j|^2$$

is valid.

We now proceed to a short description of the ideas presented in Faddeev's paper [34]. The main focus of study in [34] is the resolvent $R(\lambda) = (A - \lambda)^{-1}$ of the operator A, and the main result consists of proving its meromorphy (more precisely, meromorphy of its kernel) on the entire Riemann surface defined by the equation $s(1-s) = \lambda$. The theorem on expansion in eigenfunctions of A and the meromorphy of the function $E(z,s)$ are the consequences of this fact. We briefly outline the idea of proving meromorphy. Introduce the notation $\widetilde{R}(s) = R(\lambda)$, where $s(1-s) = \lambda$ and

$$\mathrm{Re}\, s > \tfrac{1}{2}, \quad s \bar{\in} (\tfrac{1}{2}, 1]. \tag{4.11}$$

We shall call $\widetilde{R}(s)$ the resolvent as before. In the region $\operatorname{Re} s > 1$ its kernel $r(z, z'; s)$ is given by an absolutely convergent series

$$r(z, z'; s) = \sum_{\gamma \in \Gamma} \chi(\gamma) k(z, \gamma z'; s), \tag{4.12}$$

where $k(z, z'; s)$ is Green's function for the problem $Lu - s(1-s)u = f$ on the upper half-plane H.

The problem is in extending this series to the entire s-plane as a meromorphic function. The continuation is realized by means of a Fredholm integral equation with an operator which analytically depends on the variable s. The main tool for deriving this equation is a well-known Hilbert identity for the resolvent $\widetilde{R}(s)$. Let (4.11) be fulfilled for the point s, s', then the following operator identity holds in the Hilbert space $\mathcal{H}$ (the Hilbert identity):

$$\widetilde{R}(s) - \widetilde{R}(s') = (s(1-s) - s'(1-s'))\widetilde{R}(s)\widetilde{R}(s'). \tag{4.13}$$

We now fix s' with $\operatorname{Re} s' > 1$. For the sake of brevity, we will not write the dependence on s', then we introduce the notation

$$\omega = \omega(s) = s(1-s) - s'(1-s').$$

Now (4.13) has the form

$$\widetilde{R}(s) = \widetilde{R} + \omega(s)\widetilde{R}\widetilde{R}(s). \tag{4.14}$$

This identity can be regarded as a linear equation for the operator $\widetilde{R}(s)$, taking the operator $\widetilde{R}$ to be known. But, unfortunately, (4.14) is not immediately suitable for studying solvability since the operator $\widetilde{R}$ is not compact because of the existence of the continuous spectrum of A. Nevertheless, (4.14) can be modified, if we make use of information which is contained in (4.12). In fact, that formula contains the equality $\widetilde{R} = T + V$, in which V is a compact operator and T is a bounded operator in the space $\mathcal{H}$. Besides, we can choose the operator T in such a way that the operator $I - \omega(s)T$, where I is the identity operator in $\mathcal{H}$, has an inverse operator

$$(I - \omega(s)T)^{-1} = I + \omega(s)Q(s),$$

where the bounded linear integral operator $Q(s)$ depends analytically on s, at least in the strip $0 < \operatorname{Re} s < 2$. Its kernel is given by an explicit formula. It is closely connected with Green's function of a certain ordinary differential equation. Now we modify (4.14) according to the following procedure:

$$\begin{aligned} &(I - \omega T)^{-1}\widetilde{R}(s) = \widetilde{R} + \omega V \widetilde{R}(s), \\ &\widetilde{R}(s) = (I + \omega Q(s))\widetilde{R} + \omega(I + \omega Q(s))V\widetilde{R}(s). \end{aligned} \tag{4.15}$$

Equation (4.15) is now suitable for study; however, it can be simplified by the following change of the operator variable:

$$\widetilde{R}(s) = Q(s) + (I + \omega Q(s))B(s)(I + \omega Q(s))$$

which leads to an equation for the operator $B(s)$, which is equivalent to (4.15):

$$B(s) = V + \omega V(I + \omega Q(s))B(s). \tag{4.16}$$

Solvability of (4.16) is studied not in the Hilbert space $\mathcal{H}$, but rather in a certain Banach space, since for $\operatorname{Re} s \leq \frac{1}{2}$ certain formulas mentioned above lose their meaning as operator relations in the space $\mathcal{H}$ but make sense as equalities of compositions of integral operators. In this Banach space the operator $\omega V(I+\omega Q(s))$ is compact, and it depends analytically on s, at least in the half-plane $\operatorname{Re} s > 0$. This enables us to a large extent to use the standard procedure for studying (4.16), on which we shall not dwell here.

One can prove finally that for any s there exists a unique solution $B(s)$ of (4.16) which depends analytically on s everywhere except for a discrete set of singular points. At these singular points the operator $B(s)$ has poles of finite order. The poles s_j in the half-plane $\operatorname{Re} s \geq \frac{1}{2}$ are uniquely related to the eigenvalues of the discrete spectrum of A, $s_j(1 - s_j) = \lambda_j$. It is now not difficult to extend the resolvent $\widetilde{R}(s)$ meromorphically to the entire s-plane. Then the Eisenstein–Maass series can be expressed explicitly in terms of the operator $B(s)$. One can also express the function $\varphi(s)$ of the constant term of the Eisenstein–Maass series in terms of the operator $B(s)$. Their meromorphic continuity and functional equations (4.8), (4.9) follow from here. Finally, general formula (3.6) implies the operator equality $K_{\Gamma,\chi} = \widetilde{h}(A(\Gamma;\chi))$ in the case $\chi \equiv 1$.

In conclusion, a strict statement is made of the main assertions of the chapter in the most general situation of a Fuchsian group of the first kind and its finite-dimensional singular unitary representation χ. One can find proof for these assertions in [7] (see Bibliography there).

For each cusp x_α of a normal fundamental domain F of the group Γ, more precisely, for a generator S_α of the corresponding parabolic subgroup $\Gamma_\alpha \subset \Gamma$, $\alpha = 1, \ldots, h$ (see the beginning of the section), we define the subspace V_α of the space V of the representation χ for the group Γ by the formula:
$V_\alpha = \{v \in V \mid \chi(S_\alpha)v = v\}$. We let P_α denote the orthogonal projection of V onto V_α; $\dim V_\alpha = k_\alpha$. We set $k = k(\Gamma;\chi) = \sum_{\alpha=1}^{h} k_\alpha$, where k is the full degree of singularity of χ relative to Γ.

The general Eisenstein–Maass series is defined by the formula

$$E_{\alpha l}(z,s) = E_{\alpha l}(z,s;\Gamma,\chi) = \sum_{\sigma\in\Gamma_\alpha\backslash\Gamma} y^s(\sigma_\alpha^{-1}\sigma z)\chi^*(\sigma)e_l(\alpha),$$

where $y(z) = \operatorname{Im} z$, σ_α is as at the beginning of the section, $e_l(\alpha)$ $(l = 1, \ldots, k_\alpha)$ is an element of the orthonormal basis in V_α, $*$ denotes the adjoint operator in V. The series converges absolutely in the region $\operatorname{Re} s > 1$ and in that region is an analytic function of s. As a function of z, $E_{\alpha l}(z,s) \in C^\infty(H)$; in addition, $E_{\alpha l}(\gamma z,s) = \chi(\gamma)E_{\alpha l}(z,s)$ and $LE_{\alpha l}(z,s) = s(1-s)E_{\alpha l}(z,s)$.

The Eisenstein–Maass series has the following Fourier expansion

$$P_\beta E_{\alpha l}(\sigma_\beta z, s) = \sum_{m\in\mathbf{Z}} a_{\alpha l,\beta}^{(m)}(y,s)\exp(2\pi i m x), \quad x = \operatorname{Re} z.$$

For $\operatorname{Re} s > 1$ the coefficients $a_{\alpha l,\beta}^{(m)}$ are expressed in terms of special functions and certain Dirichlet series (see [7]); when χ is one-dimensional, see also Chapter 5. The constant term $a_{\alpha l,\beta}^{(0)}(y,s)$ is equal to

$$\delta_{\alpha\beta} y^s e_l(\alpha) + \varphi_{\alpha l,\beta}(s) y^{1-s}, \tag{4.17}$$

where $\delta_{\alpha\beta}$ is the Kronecker symbol.

The elements of automorphic scattering matrix $C(s) = \{C_{d,b}(s)\}_{d,\beta=1}^{k(\Gamma;\chi)}$ are the following inner products:

$$C_{d,b}(s) = C_{\alpha l,\beta k}(s) = \langle e_k(\beta), \varphi_{\alpha l,\beta}(s)\rangle_V, \quad d = k_1 + \cdots + k_{\alpha-1} + l,$$
$$b = k_1 + \cdots + k_{\beta-1} + k, \quad 1 \le \alpha, \beta \le h, \quad 1 \le l \le k_\alpha, \quad 1 \le k \le k_\beta.$$

THEOREM 4.1. (1) *The functions $C_{\alpha l,\beta k}(s)$, $E_{\alpha l}(z,s)$ are meromorphic on the entire s-plane with order of meromorphy no greater than four*; (2) *in the half-plane $\operatorname{Re} s \ge \frac{1}{2}$ these functions have only a finite number of simple poles which lie on the interval $(\frac{1}{2}, 1]$; each such pole s_j is related to the eigenvalue λ_j of the operator $A(\Gamma;\chi)$, $s_j(1-s_j) = \lambda_j$.*

(3) *the following functional relations*

$$C_{d,\beta}(s) = \overline{C_{b,d}(s)}, \quad C(s)\cdot C(1-s) = E$$

are valid, where E is the identity matrix,

$$E_{\alpha l}(z,s) = \sum_{\beta=1}^{h}\sum_{k=1}^{k_\beta} C_{\alpha l,\beta k}(s) E_{\beta k}(z, 1-s).$$

The kernel of the resolvent of the automorphic Laplacian $A(\Gamma;\chi)$ is defined far from the spectrum by formula (4.12).

THEOREM 4.2. (1) *The kernel of the resolvent $r(z,z';s)$ for fixed z, z' with $z \ne z'$ modulo Γ-equivalence is a meromorphic function on the entire s-plane. Its poles s_j with $\operatorname{Re} s_j \ge \frac{1}{2}$ are all on the line $\operatorname{Re} s = \frac{1}{2}$ or on the interval $\frac{1}{2} < s \le 1$ of the real axis. They are all simple poles. Each such pole s_j is related to the eigenvalue λ_j of the operator $A(\Gamma;\chi)$, $s_j(1-s_j) = \lambda_j$.*

(2) *The following functional relations are valid:*

$$r^*(z,z';\bar{s}) = r(z,z';s)$$

$$r(z,z';s) - r(z,z';1-s) = \frac{1}{2s-1}\sum_{\alpha=1}^{h}\sum_{l=1}^{k_\alpha} E_{\alpha l}(z,s)\otimes E_{\alpha l}(z',1-s),$$

$*$ *denotes the adjoint operator in V, and the bar denotes complex conjugation.*

For every function $\psi \in C_0^\infty([0,\infty))$ and $e_\alpha \in V_\alpha$, $l = 1, \ldots, k_\alpha$, we define the incomplete theta-series

$$\theta_{\psi,\alpha l}(z) = \sum_{\sigma \in \Gamma_\alpha \setminus \Gamma} \psi(y(\sigma_\alpha^{-1}\sigma z))\chi^*(\sigma)e_l(\alpha)$$

(see the notation above). We let $\Theta = \Theta(\Gamma;\chi)$ denote the closed subspace in $\mathcal{H}$ spanned by all incomplete theta-series. We now introduce the orthogonal complement $\mathcal{H}_0$ of Θ in $\mathcal{H}$, $\mathcal{H}_0 = \mathcal{H} \ominus \Theta$, which is the subspace of parabolic forms of weight zero. The continuous functions $f \in \mathcal{H}_0$ (forming a dense subset in $\mathcal{H}_0$), iff the equality

$$\int_0^1 \langle f(\sigma_\alpha z), v\rangle_V \, dx = 0$$

is valid identically in $y > 0$ for any $\alpha = 1, \ldots, h$ and $v \in V_\alpha$.

Let $\Theta_0 = \Theta_0(\Gamma;\chi)$ be a linear space of the residues of all Eisenstein–Maass series at the poles lying on the interval $(\frac{1}{2}, 1]$. The inclusion $\Theta_0 \subset \Theta$ is valid. We set $\Theta_1 = \Theta \ominus \Theta_0$. We let $L_2(\mathbb{R}, \mathbb{C}^k; dt)$, where $k = k(\Gamma;\chi)$ is the full degree of singularity of the representation χ for the group Γ, denote the Hilbert space of mappings $f : \mathbb{R} \to \mathbb{C}^k$ square integrable in the Lebesgue measure dt on $\mathbb{R}$. We define the mapping $U : \mathcal{H}(\Gamma;\chi) \to L_2(\mathbb{R}, \mathbb{C}^k; dt)$, $f \to \xi = (\xi_1, \ldots, \xi_k)$ by the formula

$$\xi_{al}(t) = \int_F \langle f(z), E_{\alpha l}(z, \tfrac{1}{2} + it)\rangle_V \, d\mu(z). \tag{4.18}$$

THEOREM 4.3. (1) *The spectrum of the operator $A(\Gamma;\chi)$ consists of an absolutely continuous $k(\Gamma,\chi)$-multiple part located on the semi-axis $[\frac{1}{4},\infty)$ and a discrete part located on the semi-axis $[0,\infty)$.*

(2) *The decomposition of $\mathcal{H}$ into a direct sum of orthogonal subspaces $\mathcal{H} = \mathcal{H}_0 \oplus \Theta_0 \oplus \Theta_1$ holds, and othogonal projections on each of these subspaces commute with $A(\Gamma;\chi)$; $\mathcal{H}_0 \oplus \Theta_0$ is the subspace of the discrete spectrum, Θ_1 is the subspace of the continuous spectrum of $A(\Gamma;\chi)$, and the subspace Θ_0 is finite-dimensional.*

(3) *If P is the orthogonal projection in $\mathcal{H}$ on $\mathcal{H}_0 \oplus \Theta_0$ and U is the operator defined by formula (4.18), then $U^*U = I - P$, where U^* is the adjoint operator, I is the identity operator in $\mathcal{H}$.*

(4) *Let k and h be connected by formulas (4.4) and the operator $K_{\Gamma,\chi}$ be defined by the kernel (4.5), then the following operator equality and equality of kernels hold:*

$$\widetilde{h}(A(\Gamma;\chi)) = K_{\Gamma,\chi}$$

$$k_{\Gamma,\chi}(z,z') = \sum_j \widetilde{h}(\lambda_j)\overline{v_j(z)} \otimes v_j(z') +$$

$$+ \frac{1}{4\pi}\sum_{\alpha=1}^{h}\sum_{l=1}^{k_\alpha}\int_{-\infty}^{\infty} \widetilde{h}(\tfrac{1}{4} + r^2)E_{\alpha l}(z, \tfrac{1}{2} + ir) \otimes \overline{E_{\alpha l}(z', \tfrac{1}{2} + ir)} \, dr,$$

where $\{v_j\}$ *is a basis of eigenfunctions of* $A(\Gamma;\chi)$ *in* $\mathcal{H}_0 \oplus \Theta_0$, $Av_j = \lambda_j v_j$, *the integral and series converge in the metric of the space* $\mathcal{H}$.

CHAPTER 5

Harmonic Analysis of Automorphic Functions. Estimates for Fourier Coefficients of Parabolic Forms of Weight Zero

The aims of this chapter are to assess known explicit formulas for the Fourier coefficients of automorphic functions playing an important role in spectral theory, and transferring certain classical estimates from the theory of analytic modular forms to non-analytic parabolic forms of weight zero.

In classical theory of automorphic functions and forms and also in its applications the Fourier expansions play a significant role. It suffices to say that the first examples of analytic automorphic forms were constructed as generating functions for number sequences, interesting from the viewpoint of arithmetics and number theory. The classical example (see Chapter 2) is the theta-series $\theta(z;Q)$ generated by a positive quadratic form Q with integer coefficients

$$\theta(z;Q) = \sum_M e(zQ(M)), \tag{5.1}$$

where M runs through the set of integer vectors. It is an analytic automorphic form of weight k relative to the group $\Gamma_0(N)$ with the character χ, i.e., in particular, it satisfies the functional equation

$$\theta\Big(\frac{az+b}{cz+d};Q\Big) = (cz+d)^k\chi(\gamma)\theta(z;Q), \quad \gamma = \begin{pmatrix} a & b \\ c & d \end{pmatrix} \in \Gamma_0(N).$$

In these formulas $\Gamma_0(N)$ is the congruence subgroup of Hecke (see Chapter 8 and later in this chapter), the parameters N, k, χ are defined by the parameters of Q. The series (5.1) can be considered as the Fourier series

$$\theta(z;Q) = \sum_{n=0}^{\infty} a_n(Q)e(nz), \tag{5.2}$$

where $a_n(Q)$ is the number of representations of n by the quadratic form. Studying analytic properties of the function $\theta(z;Q)$ (as a function of z) enables us to understand the properties of the numbers $a_n(Q)$, and in particular, their asymptotic behaviour as $n \to \infty$.

On the contrary, if there is an automorphic function or form defined in any way, then its sequence of Fourier coefficients is very special, and its properties are very important for the global number theory, and in particular, for understanding general reciprocity laws according to Langlands (for details see Chapter 12).

The properties of automorphy and analyticity enable one to obtain, in a certain sense, the formulas for the Fourier coefficients of a given form or function. Although the mentioned formulas give, as usual, no essential information about the coefficients, and so are illusory in this sense, nevertheless they turn out to be occasionally useful (see Chapter 13).

It is no wonder that Fourier expansions were considered in the first papers on spectral theory of automorphic functions. We keep in mind here the fundamental papers by Maass [130] and Selberg [163]. We now present one of the results. Let Γ be, as before, a Fuchsian group of the first kind with the fundamental domain F, χ be the one-dimensional unitary representation of the group Γ. Later on in this chapter we shall suppose that Γ is such that F is non-compact. (For the case of the compact F, see Chapter 13.)

Let Γ_α be a parabolic subgroup of Γ which stabilizes the cusp α of F and $\sigma_\alpha \in \mathrm{PSL}(2,\mathbb{R})$ be such that

$$\sigma_\alpha\infty = \alpha, \quad \sigma_\alpha^{-1}\Gamma_\alpha\sigma_\alpha = \Delta = \{\gamma^n \mid n \in \mathbb{Z}\}, \quad \gamma = \begin{pmatrix} 1 & 1 \\ 0 & 1 \end{pmatrix} \ (\mathrm{mod} \pm 1). \quad (5.3)$$

We define $\xi \in \mathbb{R}$ by conditions

$$\chi(\sigma_\alpha\gamma\sigma_\alpha^{-1}) = e(\xi), \quad 0 \le \xi \le 1. \quad (5.4)$$

Suppose that the function $u : H \to \mathbb{C}$ has the property

$$u(gz) = \chi(g)u(z), \quad g \in \Gamma, \quad z \in H. \quad (5.5)$$

Then the following equality holds

$$u(\sigma_\alpha(z+1)) = e(\xi)u(\sigma_\alpha z), \quad z \in H \quad (5.6)$$

and, consequently, the following Fourier expansion exists

$$u(\sigma_\alpha z) = \sum_{m-\xi\in\mathbb{Z}} c_{\alpha,m}(y)e(mx), \quad z \in H, \quad x = \mathrm{Re}\, z, \quad y = \mathrm{Im}\, z, \quad (5.7)$$

if to require any conditions for u guaranteeing the convergence of the series (5.7) to the expanded function (the summation in (5.7) is over all of $m \in \mathbb{R}$ for which $m - \xi \in \mathbb{Z}$).

We now suppose in addition to (5.5) that u satisfies the following differential equation with the Laplace–Beltrami operator L; $\lambda \in \mathbb{C}$:

$$Lu = \lambda u. \quad (5.8)$$

Then the equality (5.7) holds. We may study the coefficients in (5.7) using the method of separation of variables. We substitute (5.7) in (5.8) and equate the coefficients with equal exponents. We obtain the equation

$$\frac{\mathrm{d}}{\mathrm{d}y}c_{\alpha,m}(y)+\left(\frac{\lambda}{y^2}-4\pi^2m^2\right)c_{\alpha,m}(y)=0. \tag{5.9}$$

Equation (5.9) can easily be reduced to the Bessel equation. Thus we arrive at the following theorem according to Maass [130].

THEOREM 5.1. *Let $u: H\to\mathbb{C}$ be a function satisfying* (5.5), (5.8) *with Γ, χ, α, σ_α, ξ as above. Then the following equality is valid:*

$$\begin{aligned}u(\sigma_\alpha z)&=\rho(0,\alpha)y^{1/2+i\kappa}+\tilde{\rho}(0,\alpha)y^{1/2-i\kappa}+\sum_{m-\xi\in\mathbb{Z}}\{\rho(m,\alpha)\sqrt{y}K_{i\kappa}(2\pi|m|y)+\\&+\tilde{\rho}(m,\alpha)\sqrt{y}I_{i\kappa}(2\pi|m|y)\}e(mx),\end{aligned} \tag{5.10}$$

$$z\in H,\quad x=\operatorname{Re}z,\quad y=\operatorname{Im}z;$$
$$\kappa\in\mathbb{R}_+\cup i\mathbb{R}_+,\quad \kappa^2=\lambda-\tfrac14;\quad \rho(m,\alpha),\ \tilde{\rho}(m,\alpha)\in\mathbb{C}.$$

What is more, if $u\in\mathcal{H}$ (see Chapter 4), then

$$\begin{aligned}&(1)\quad \tilde{\rho}(0,\alpha)=0,\quad \text{for all of } m;\\&(2)\quad \rho(0,\alpha)=0,\quad \text{if}\quad \lambda\ge\tfrac14\end{aligned} \tag{5.11}$$

We clarify how to obtain the last assertion of the theorem. If $u\in H$, then the following formula is valid

$$(u,u)=\int_F|u(z)|^2\,\mathrm{d}\mu(z)=\int_{\sigma_\alpha^{-1}F}|u(\sigma_\alpha z)|^2\,\mathrm{d}\mu(z)<\infty; \tag{5.12}$$

after substituting (5.10) in (5.12), we easily arrive at (5.11).

For the Eisenstein–Maass series defined in Chapter 4, and for certain other functions, the coefficients of Fourier expansion can be calculated explicitly. The kernel of the resolvent of the automorphic Laplacian is among those functions.

Let us first consider the Eisenstein–Maass series. Let β be one of the parabolic vertices of F, and let χ be singular at β, i.e. $\chi(\Gamma_\beta)=1$; E_β be the Eisenstein–Maass series corresponding to this vertex (see Chapter 4): $E_\beta(z;s;\chi)=E(z;s;\beta;\Gamma;\chi)$ from Chapter 4.

THEOREM 5.2. *The following expansion holds:*

$$E_\beta(\sigma_\alpha z;s;\chi)=\sum_{m-\xi\in\mathbb{Z}}a_{\beta\alpha,m}(y;s;\chi)e(mx), \tag{5.13}$$

$z\in H$, $x=\operatorname{Re}z$, $y=\operatorname{Im}z$. *In this expansion*

$$a_{\beta\alpha,0}(y;s;\chi)=\delta_{\alpha\beta}y^s+y^{1-s}\varphi_{\beta\alpha}(s);$$
$$a_{\beta\alpha,\xi}(y;s;\chi)=0,\quad \xi\ne0;$$
$$\varphi_{\beta\alpha}(s)=\pi^{1/2}\frac{\Gamma(s-\frac12)}{\Gamma(s)}\varphi_{\beta\alpha,0}(s);\quad \varphi_{\beta\alpha}(s)=C_{\beta1,\alpha1}(s)\ \textit{from Chapter 4}$$

$$(\dim V_\alpha = \dim V_\beta = 1);$$

$$a_{\beta\alpha,m}(y;s;\chi) = 2\pi^s|m|^{s-1/2}\Gamma(s)^{-1}y^{1/2}K_{s-1/2}(2\pi|m|y)\varphi_{\beta\alpha,m}(s),$$

$$m-\xi\in\mathbb{Z},\quad m\neq 0. \tag{5.14}$$

For $\operatorname{Re} s > 1$ *the function* $\varphi_{\beta\alpha,m}(s)$ *is given by an absolutely convergent Dirichlet series*

$$\varphi_{\alpha\beta,m}(s) = \sum_c \frac{1}{|c|^{2s}}\left(\sum_d \chi_{\beta\alpha}\left(\begin{pmatrix} * & * \\ c & d\end{pmatrix}\right)e\left(m\tfrac{d}{c}\right)\right),$$

the summation is taken over c, d *satisfying conditions:*

$$c>0;\quad 0\le d<c;$$

there exist a *and* b *such that* $\begin{pmatrix} a & b \\ c & d\end{pmatrix} \in \sigma_\beta^{-1}\Gamma\sigma_\alpha$; *here* $\chi_{\beta\alpha}(h) = \chi(\sigma_\beta h\sigma_\alpha^{-1})$ *for* $h\in\sigma_\beta^{-1}\Gamma\sigma_\alpha$ *(see the list of notations).*

Theorem 5.2 follows from Theorem 5.1, the definition of Eisenstein–Maass series and the estimate in the region $\operatorname{Re} s > 1$

$$E_\beta(\sigma_\alpha z;s;\chi) = O(y^{\max\{\operatorname{Re} s,\,1-\operatorname{Re} s\}}),\quad y\to\infty,$$

which can be easily proved.

The main application of Theorem 5.2 at present is as follows: the Eisenstein–Maass series can be extended meromorphically in s onto all of $\mathbb{C}$. This assertion is proved by the methods of spectral theory of automorphic functions (see Chapter 4). Thus, all Dirichlet series can be extended meromorphically onto all of $\mathbb{C}$, which is very significant for number theory. (For such applications, see Chapter 10.)

Consider now the kernel of the resolvent of the automorphic Laplacian (see Chapter 4), and the Fourier expansion for it. For simplicity, we restrict ourselves to the case where Γ contains a transformation $z\to z+1$, in this case $\sigma_\alpha = 1$. Since the following equality is valid

$$r(gz,g'z';s;\Gamma;\chi) = \chi(g)\chi^{-1}(g')r(z,z';\Gamma;\chi)$$

for all $z,z'\in H$, $g,g'\in\Gamma$, then one can expand the function $z'\to r(z,z';s;\Gamma;\chi)$ in a Fourier expansion of the form (5.10) in powers of $e(x')$, $x' = \operatorname{Re} z'$, and then expand every coefficient of this Fourier series in powers of $e(x)$, $x = \operatorname{Re} z$. To state the result, we have to define the sums of Kloosterman $S(n,m,c)$. Let c be such that

$$\begin{pmatrix} a & b \\ c & d\end{pmatrix}\in\Gamma \tag{5.15}$$

for certain a, b, d and let $m-\xi$, $n-\xi \in \mathbb{Z}$. Then we set

$$S(n,m,c)=\sum_{0\le d<c}\overline{\chi}\left(\begin{pmatrix} a & b\\ c & d\end{pmatrix}\right)e\left(\frac{ma+nd}{c}\right), \tag{5.16}$$

the summation is taken over all d, $0\le d<c$, for which a,b exist with the condition (5.15) (the sum does not depend on the choice of a,b).

THEOREM 5.3. *The following Fourier expansion holds*

$$\begin{aligned} r(z,z';s;\Gamma;\chi) &= \delta_{\xi 0}\frac{y'^{\,1-s}}{1-2s}E_\infty(z;s;\chi)+\\ &+\sum_{\substack{m-\xi\in\mathbb{Z}\\ m\neq 0}}\hat{F}_m(z,s)(4|m|y')^{1/2}K_{s-1/2}(2\pi|m|y')e(mx'); \end{aligned} \tag{5.17}$$

where $z,z'\in H$, $x=\operatorname{Re} z$, $x'=\operatorname{Re} z'$, $y=\operatorname{Im} z$, $y'=\operatorname{Im} z'$, *and we suppose that* $y'>y$. *The coefficients* F_m *are defined by the following equality*

$$\begin{aligned} \hat{F}_m(z,s) &= -\frac{4^{s-1}\pi^{1/2}\Gamma(s)\Gamma(s-\frac12)}{|m|^{1/2}\Gamma(2s)}+\\ &+\sum_{\gamma\in\Gamma_\infty\backslash\Gamma}\overline{\chi}(\gamma)(\operatorname{Im}\gamma z)^{1/2}I_{s-1/2}(2\pi|m|\operatorname{Im}\gamma z)e(-m\operatorname{Re}\gamma z). \end{aligned} \tag{5.18}$$

The series converges absolutely in the region $\operatorname{Re} s>1$.

THEOREM 5.4 *The following expansion holds for coefficients* $\hat{F}_m$ *from Theorem 5.3* ($\operatorname{Re} s>1$):

$$\begin{aligned} -\hat{F}_m(z,s) &= \frac{4^{s-1}\Gamma(s+\frac12)\Gamma(s)}{\sqrt{\pi|m|}\,\Gamma(2s)}y^{1/2}I_{s-1/2}(2\pi|m|y)e(-mx)+\\ &+\sum_{\substack{n-\xi\in\mathbb{Z}\\ n\neq 0}}\left[\frac{1}{2\sqrt{|mn|}}\sum_{c>0}\frac{S(-m,n,c)}{c}\Omega\left(\frac{4\pi\sqrt{|mn|}}{c}\right)\right]\times\\ &\times(4|n|y)^{1/2}K_{s-1/2}(2\pi|n|y)e(nx); \end{aligned} \tag{5.19}$$

here

$$\Omega=\begin{cases} J_{2s-1}, & if \quad mn<0,\\ I_{2s-1}, & if \quad mn>0,\end{cases}$$

$z\in H$, $y=\operatorname{Im} z$, $x=\operatorname{Re} z$ (*see the list of notations*).

In the proof of Theorems 5.3 and 5.4, one makes use of the method of constructing Green's function for the equation with separable variables, and the explicit form of the normed Siegel–Selberg series $\hat{F}_m(z,s)$, as well.

We now proceed to the question of the Fourier coefficients of parabolic forms of weight zero which are in the expansion in eigenfunctions of an automorphic Laplacian (see Chapter 4). This is perhaps the right place to state the following:

these parabolic forms are objects so mysterious that at present there are practically no explicit formulas for their Fourier coefficients. In the few cases where we know indirectly of their existence (see Chapter 8), the question can only be of one or another estimate for them.

The following general theorem is by the method of proving and by precision the analog of Hardy's estimate known in the theory of analytic automorphic forms.

THEOREM 5.5. *Let the function $u \in \mathcal{H}$ satisfy the properties $Lu = \lambda u$, $\lambda \in \mathbb{R}$, u is bounded such that $c = \max_{z\in F} |u(z)|$ is defined. Then in the notation of Theorem 5.1 we have*

$$\rho(m, \alpha) \ll |m|^{1/2}, \quad m \neq 0, \tag{5.20}$$

where the constant in the symbol $\ll$ depends only on λ and c.

The proof of the theorem is very simple. From formula (5.10) we have the estimate

$$|\rho(m, \alpha) y^{1/2} K_{i\kappa}(2\pi|m|y)| = \left| \int_0^1 u(\sigma_\alpha z) e(-mx)\, dx \right| \leq c$$

for any $y \in \mathbb{R}_+$. So the following inequality is valid:

$$|\rho(m, \alpha)| \leq c \left\{ \max_{y\in \mathbf{R}} y^{1/2} K_{i\kappa}(2\pi|m|y) \right\}^{-1} = cc'|m|^{1/2},$$

if to set

$$c' = \max_{t\in \mathbf{R}_+} \left(\frac{t}{2\pi}\right)^{1/2} K_{i\kappa}(t);$$

the proof is complete.

Consider now the less general situation. We recall a classical definition

$$\Gamma_0(N) = \left\{ \begin{pmatrix} a & b \\ c & d \end{pmatrix} \in \mathrm{PSL}(2, \mathbb{Z}) \middle| c \equiv 0 (\mathrm{mod\ N}) \right\}$$

of Hecke's congruence subgroup of a modulo N in the modular group $\mathrm{PSL}(2, \mathbb{Z})$, $N \in \mathbb{Z}_+$. $\Gamma_0(N)$ is an arithmetical Fuchsian group of the first kind with a non-compact fundamental domain (for more detailed information on arithmetical groups see Chapter 8). We let $\hat{\chi}$ denote a Dirichlet character modulo N, $\hat{\chi}(-1) = 1$. We remind the definition of a general character mod N. This number-theoretic function given on the set of integers, is defined by the following four properties: (1) $\hat{\chi}(a) = 0$ for $(a, N) \neq 1$; (2) $\hat{\chi}(1) \neq 0$; (3) $\hat{\chi}(ab) = \hat{\chi}(a)\hat{\chi}(b)$; (4) $\hat{\chi}(a) = \hat{\chi}(b)$, if $a \equiv b (\mathrm{mod}\, N)$. The chosen Dirichlet character with condition $\hat{\chi}(-1) = 1$ defines correctly a one-dimensional unitary representation χ of the group $\Gamma_0(N)$ by the following formula:

$$\chi(\gamma) = \hat{\chi}(d), \quad \gamma = \begin{pmatrix} a & b \\ c & d \end{pmatrix} \in \Gamma_0(N). \tag{5.21}$$

In the sequel up to the end of the chapter, unless otherwise stated, we let Γ denote any group $\Gamma_0(N)$, and χ denote the representation from (5.21). In this case the subspace of parabolic forms of weight zero is non-trivial, and Weyl's law (see Chapter 8) is valid for the distribution function for eigenvalues of the discrete spectrum of the automorphic Laplacian.

The main hypothesis of this circle of problems consists of proving the estimate

$$\rho(m,\alpha) \ll d(m) \tag{5.22}$$

for the Fourier coefficients of the expansion (5.10) for parabolic form $u \in \mathcal{H}(\Gamma;\chi)$ of weight zero. In (5.22) $m \in \mathbb{Z}$, $m \neq 0$, $d(m)$ is the number of divisors of m, the constant in the symbol $\ll$ depends only on the function u; in addition, the numbers m and N are supposed to be relatively prime, i.e. $(m,N)=1$.

The supposition (5.22) is the analog of the classical hypothesis of Ramanujan–Petersson on the estimate for Fourier coefficients of analytic parabolic forms proved by P. Deligne (see [66]), and it seems to be very natural. We can indicate at least five authors who at different times and in different terms have proposed hypotheses similar to (5.22) and have made progress in proving them. Those authors are I. Satake, I. I. Pjatetskiĭ–Shapiro, N. V. Kuznetsov, R. W. Bruggeman, N. V. Proskurin (see [160], [15], [18], [56], [29]).

Except for the limited interest for the theory of parabolic forms of weight zero, the hypothesis (5.22) is significant for number theory. Its proof implies considerable progress in uniform estimates of Kloosterman sums, and, in particular, makes Linnik's hypothesis (see Chapter 13) more convincing.

The hypothesis can be reformulated in terms of eigenvalues of Hecke operators. The Hecke operators, for example, as operators in the Hilbert space of automorphic functions $\mathcal{H}(\Gamma;\chi)$ are defined by the following equality

$$(T_m u)(z) = \frac{1}{m^{1/2}} \sum_{ad=m} \sum_{b (\mathrm{mod}\, d)} \overline{\chi}(d) u\Big(\frac{az+b}{d}\Big), \quad m \in \mathbb{Z}_+, \quad (m,N)=1, \tag{5.23}$$

where $u \in \mathcal{H}(\Gamma;\chi)$, the summation is taken over integers a, b, d with conditions given above, the bar means the complex conjugation (for the definition and main properties of Hecke operators in other situations, see Chapter 8). It is known that the Hecke operators commute with one another and are permutable with the automorpic Laplacian $A(\Gamma;\chi)$. In addition, they map the subspace of parabolic forms of weight zero $\mathcal{H}_0(\Gamma;\chi)$ into itself. There exists an orthonormal basis in the subspace $\mathcal{H}_0(\Gamma;\chi)$ consisting of common eigenfunctions of T_m and $A(\Gamma;\chi)$. Let u be one of the functions of this basis, then we have

$$A(\Gamma;\chi) = \lambda u$$
$$T_m u = \tau(m) u.$$

It is easy to prove (substituting (5.10) with $\alpha = \infty$ into (5.23)) that the equality

$$\sum_{d|(n,m)} \chi(d)\rho\left(\frac{nm}{d^2}, \infty\right) = \tau(m)\rho(n, \infty), \tag{5.24}$$

holds for all $m \in \mathbb{Z}$, $m \neq 0$. In particular,

$$\frac{\rho(mn, \infty)}{\rho(n, \infty)} = \tau(m) \tag{5.25}$$

is valid, if $(m, n) = 1$ and $\rho(n, \infty) \neq 0$.

Now, if we suppose the validity of (5.22) (with $\alpha = \infty$), then, taking into account (5.24), (5.25), one can prove the following inequality

$$|\tau(p)| \leq 2 \tag{5.26}$$

for each prime p, $(p, N) = 1$.

Unfortunately, neither the hypothesis (5.22), nor the inequality (5.26) have yet been proven. It is known only that the following two inequalities hold:

$$|\tau(p)| \leq p^{1/4} + p^{-1/4}, \tag{5.27}$$

$$\rho(m, \alpha) \ll d(m)m^{1/4}. \tag{5.28}$$

The proof of (5.27), (5.28) is based on investigations of the real-analytic Poincaré series first considered by Selberg (see [166]). The real-analytic Poincaré series is defined for each $m \in \mathbb{Z}$ by the following equality:

$$U_m(z, s) = \sum_{\sigma \in \Gamma_\infty \backslash \Gamma} \overline{\chi}(\sigma) y(\sigma z)^s e((m - \xi)\sigma z), \quad z \in H. \tag{5.29}$$

The series converges absolutely and uniformly on compact sets in the domain $\operatorname{Re} s > 1$ and $U_m(\cdot, s) \in \mathcal{H}(\Gamma; \chi)$. For functions (5.29) the Fourier coefficients are computed in the same way, as the Fourier coefficients of Eisenstein–Maass series. Here, instead of the Dirichlet series (5.14), the series containing the Kloosterman sums (5.16) appears.

The function $U_m(\cdot, s)$ can be expanded into a series in eigenfunctions of the automorphic Laplacian. In this expansion, the coefficients are meromorphic functions in s. In such a way Selberg extended the real-analytic Poincaré series meromorphically onto the entire s-plane.

Kuznetsov (see [18], [19]) found a very convenient method of applying the series (5.29) to study Fourier coefficients of parabolic forms. We outline here the principal idea behind the method. He considered the inner product of two series

$$(U_m(\cdot, s_1), U_n(\cdot, s_2)) \tag{5.30}$$

in the Hilbert space $\mathcal{H}(\Gamma; \chi)$. The inner product (5.30) can be computed firstly by means of the Parseval equality for the expansion in eigenfunctions of the automorphic Laplacian (see Chapter 4). On the other hand to compute the inner product

(5.30) on can use the Fourier expansion for $U_n(\cdot, s_1)$ and the definition (5.29). In both cases the calculations begin with the equality

$$\int_F U_m(z,s)\overline{f(z)}\,\mathrm{d}\mu(z) = \int_0^\infty \int_0^1 (\operatorname{Im} z)^s e(mz)\overline{f(z)}\,\mathrm{d}\mu(z), \tag{5.31}$$

which is valid for any function $f \in \mathcal{H}(\Gamma;\chi)$, and also in the case, when f is the Eisenstein–Maass series. Then we substitute in (5.31) instead of f its Fourier expansion and integrate it term by term. So we obtain two expressions for the inner product (5.30). After comparing them, Kuznetsov arrived at the following two theorems (see [19]). In order to state them we introduce some notation. Let $\{v_j\}_{j\geq 1}$ be an orthonormal basis of the subspace of the discrete spectrum $\mathcal{H}_0(\Gamma;\chi)\oplus\theta_0(\Gamma;\chi)$ of the automorphic Laplacian $A(\Gamma;\chi)$ (see Chapter 4) formed by its eigenfunctions; let $\lambda_1 \leq \lambda_2 \leq \cdots$ be corresponding eigenvalues, $Av_j = \lambda_j v_j$, and let

$$v_j(z) = \sum_{m\in\mathbb{Z}} \rho_j(m) y^{1/2} K_{i\kappa_j}(2\pi|m|y)e(mx)$$

be the Fourier expansion ($\rho_j(m)$ and κ_j be $\rho(m,\infty)$ and κ from Theorem 5.1 respectively for $u = v_j$).

THEOREM 5.6. *Let a function $\varphi : [0,\infty) \to \mathbb{C}$ be triply continuously differentiable and satisfy the conditions*

$$\varphi(0) = \varphi'(0) = 0; \quad \varphi(x) \ll x^{-1-\epsilon},$$
$$\varphi'(x),\ \varphi''(x),\ \varphi'''(x) \ll x^{-2-\epsilon}, \quad x \to \infty,$$

for a certain $\epsilon > 0$. Then for $m, n \in \mathbb{Z}_+$ the following equality holds:

$$\sum_{c>0} \frac{S(n,m,c)}{c}\varphi_*\left(\frac{4\pi\sqrt{mn}}{c}\right)+$$
$$+\frac{\delta_{nm}}{2\pi}\int_0^\infty J_0(x)\varphi(x)\,\mathrm{d}x = \sum_{j=1}^\infty \frac{\overline{\rho_j(n)}\rho_j(m)}{\operatorname{ch}\pi\kappa_j}\hat{\varphi}(\kappa_j)+$$
$$+2/\pi\sum_\alpha \int_0^\infty \left(\frac{m}{n}\right)^{ir}\overline{\varphi_{\alpha\infty,n}(\tfrac12+ir)}\varphi_{\alpha\infty,m}(\tfrac12+ir)\hat{\varphi}(r)\,\mathrm{d}r.$$

(The Kuznetsov summation formula.)

Here

$$\hat{\varphi}(r) = \frac{\pi i}{2\operatorname{sh}\pi r}\int_0^\infty \{J_{2ir}(x) - J_{-2ir}(x)\}\varphi(x)\frac{\mathrm{d}x}{x},$$
$$\varphi_*(x) = \varphi(x) - \sum_{k=0}^\infty 2(2k+1)J_{2k+1}(x)\int_0^\infty J_{2ir}(y)\varphi(y)\frac{\mathrm{d}y}{y};$$

the summation in the left-hand side is over all $c > 0$ such that there exist a, b, d

with

$$\begin{pmatrix} a & b \\ c & d \end{pmatrix} \in \Gamma;$$

the summation over α in the right-hand side is extended onto all cusps of the domain F, at which χ is singular.

THEOREM 5.7. *If $\Gamma = \mathrm{SL}_2(\mathbb{Z})$, $\chi = 1$, then the following equality holds:*

$$\sum_{j:\kappa_j \le X} \frac{|\rho_j(n)|^2}{\operatorname{ch} \pi\kappa_j} = \frac{1}{\pi^2} X^2 + O(X \log X - Xn^{\epsilon} - n^{1/2+\epsilon}), \quad X \ge 2,$$

with the constant in the symbol O depending only on $\epsilon > 0$.

In proving Theorem (5.7), one makes use of Weil's estimate for the Kloosterman sums (see [182]). To be precise, it should be pointed out that Theorem 5.6 was proved by Kuznetsov for the case of the modular group $\mathrm{SL}_2(\mathbb{Z})$ and the trivial representation χ, and it was generalized by Proskurin later on (see [30], [31]). Theorem 5.7 can also be generalized to the case $\Gamma = \Gamma_0(N)$ (see Chapter 13).

Theorems 5.6, 5.7 imply the desired inequalities (5.27), (5.28) (see [18]).

Theorem 5.7 enables one to obtain, for example, the following assertion for the modular group Γ and the trivial representation χ: the estimate (5.26) holds for 'almost all' functions v_j. More precisely, supposing $T_p v_j = \tau_j(p) v_j$, then among the functions v_j, for which the inequality $j \le X$ is valid, there exist no more than $o(X)$ functions such that (5.26) may not be valid (with $\tau_j(p)$ instead of $\tau(p)$). We note that Weyl's law (see Chapter 8) implies that the set of all functions v_j, $j \le X$, has the order $\frac{1}{12}X$, as X approaches infinity. It was Bruggeman who first obtained this result and reported it in [56].

Recently Iwaniec and Deshouillers (see [107], [68]) obtained for Fourier coefficients of parabolic forms of weight zero the inequalities of the type 'large sieve' from analytic number theory. The simplest of their results is given by the following theorem.

THEOREM 5.8. *Under the assumptions of Theorem 5.6, for any $\epsilon \in R_+$ the following estimate holds:*

$$\sum_{\lambda_j \le X} \frac{1}{\operatorname{ch} \pi\kappa_j} \Big| \sum_{Y < n \le 2Y} a_n \rho_j(n) \Big|^2 \ll (X + N^{-1} Y^{1+\epsilon}) \sum_{Y < n \le 2Y} |a_n|^2;$$

here $X, Y \in \mathbb{R}$, $X \ge 1$, $Y \ge \frac{1}{2}$; $a_n \in \mathbb{C}$; the constant in the symbol $\ll$ depends only on ϵ.

CHAPTER 6

The Selberg Trace Formula for Fuchsian Groups of the First Kind

An expansion in eigenfunctions of the automorphic Laplacian we have discussed in Chapter 4 is the basis of the Selberg trace formula on the Lobachevsky plane. In this chapter we acquaint the reader with the ideas behind the derivation of the formula. The more general cases of the Selberg trace formula are discussed in Chapter 11.

As in chapter 4, we consider first the simple case of a co-compact discrete group Γ, then we give a short derivation of the trace formula for a model non-co-compact Fuchsian group Γ and the trivial representation χ. Finally, we state theorems describing the general case of a non-co-compact Fuchsian group Γ and its finite-dimensional unitary representation χ.

We begin the exposition with the assumption that Γ is a strictly hyperbolic group (see Chapter 4), and χ is its arbitrary finite-dimensional unitary representation. First of all, it is not difficult to prove the following assertion: if $k \in C_0^\infty([0,\infty))$, then a linear integral operator $K_{\Gamma,\chi}$ defined by its kernel (4.5) is of trace class in the space $\mathcal{H}(\Gamma;\chi)$, and the following equality is valid:

$$\int_F \operatorname{tr}_V K_{\Gamma,\chi}(z,z)\,\mathrm{d}\mu(z) = \sum_j \widetilde{h}(\lambda_j), \tag{6.1}$$

where λ_j runs through the spectrum of the operator $A(\Gamma;\chi)$ (counting the multiplicity), the function $\widetilde{h}$ is from the transformation (4.4), tr_V means the trace of the operator in the space V. (For proof that $K_{\Gamma,\chi}$ is of trace class, see Chapter 11.)

Formula (6.1) is the typical trace formula from the standpoint of the theory of operators in the Hilbert space. The Selberg trace formula contains a large body of information since the left-hand side of (6.1) can be computed effectively. The main body of these calculations is so elegant and simple that we present it here. The left-hand side of (6.1) is transformed to the form

$$\begin{aligned}\sum_{\gamma\in\Gamma} \operatorname{tr}_V \chi(\gamma) \int_F k(z,\gamma z)\,\mathrm{d}\mu(z) &= \sum_{\{\gamma\}_\Gamma} \operatorname{tr}_V \chi(\gamma) \sum_{\gamma'\in\Gamma_\gamma\backslash\Gamma} \int_F k(z,\gamma'\gamma^{-1}\gamma' z)\,\mathrm{d}\mu(z)\\ &= \sum_{\{\gamma\}_\Gamma} \operatorname{tr}_V \chi(\gamma) \int_{F_\gamma} k(z,\gamma z)\,\mathrm{d}\mu(z),\end{aligned} \tag{6.2}$$

where we use the following notation: $\{\gamma\}_\Gamma$ is the conjugacy class in Γ, containing γ, Γ_γ is the centralizer of the element γ in Γ; F_γ is the fundamental domain of the group Γ_γ on H; $\{\gamma\}_\Gamma$ runs through the set of all conjugacy classes. The transformations (6.2) are of the most general nature. They are the basis of the Selberg trace formula for any weakly symmetric Riemann space (see Chapter 12). What is their meaning? The integral in the right-hand side of formula (6.2) turns out to be explicitly computable in terms of the Fourier–Harish–Chandra transformation h of the function k. This is possible for two reasons. In the first place, the domain F_γ has a simpler form than F, because of a simpler structure of the group Γ_γ in comparison with Γ. Secondly, by means of a change in the variable $z \to gz$ with a suitable element $g \in \mathrm{PSL}(2,\mathbb{R})$ in the mentioned integral, one can transform the element γ to a canonical form which also simplifies the calculations. All these calculations can be carried out in more detail in the case we now consider. All conjugacy classes $\{\gamma\}_\Gamma$ are hyperbolic except for $\{1\}_\Gamma$, where 1 is the identity of the group. The element γ of the finite order is primitive in Γ, if it is not an essential power of another element. This is similar for the class $\{\gamma\}_\Gamma$. It is well-known that the set of all primitive hyperbolic classes in any Fuchsian group of the first kind is infinite. Then, every hyperbolic element γ is conjugate to an element (transformation) $z \to N(\gamma)z$, $N(\gamma) > 1$, ($z \in H$) in the group $\mathrm{PSL}(2,\mathbb{R})$. Following Selberg, we shall call $N(\gamma)$ the norm of the hyperbolic element γ (or the norm of the conjugacy class $\{\gamma\}_\Gamma$). Now, taking into account formula (4.4), by means of simple calculation, we arrive at the following assertions.

LEMMA 6.1. *The following equality holds;*

$$\int_F k(z,z)\,\mathrm{d}\mu(z) = \frac{|F|}{4\pi}\int_{-\infty}^{\infty} r\,\mathrm{th}(\pi r)h(r)\,\mathrm{d}r,$$

where $|F|$ is the volume of F in the measure $\mathrm{d}\mu$, $k(z,z) = k(t(z,z)) = k(0)$, th *is the hyperbolic tangent.*

LEMMA 6.2. *Let $\gamma' = \gamma^k$ be a k-*th *power of a primitive hyperbolic element $\gamma \in \Gamma$. Then*

$$\int_{F_{\gamma'}} k(z,\gamma' z)\,\mathrm{d}\mu(z) = \frac{\ln N(\gamma)}{N(\gamma)^{k/2} - N(\gamma)^{-k/2}}\, g(k \ln N(P)).$$

Now, equating the right-hand sides of (6.1), (6.2) and using Lemmas 6.1, 6.2, we arrive at the desired Selberg trace formula for the kernel of the operator $K_{\Gamma,\chi}$ with a finite function k. Since the formula obtained contains the function h instead of k, it is natural to lay certain restrictions on h. The trace formula turns out to be valid, as an identity relative to the function h, for a broader class of functions. We give the final formula as a theorem (see [164]).

THEOREM 6.1. *Suppose that the function h satisfies the conditions:*
(1) $h(r) = h(-r)$,

(2) *$h(r)$ is analytic in the strip $|\operatorname{Im} r| < \frac{1}{2} + \epsilon$, $\epsilon > 0$,*
(3) *$h(r) = O\big((1+|r|^2)^{-1-\epsilon}\big)$ in this strip.*
Then the following formula (Selberg trace formula) holds:

$$\sum_j h(r_j) = \dim V \frac{|F|}{2\pi} \int_{-\infty}^{\infty} r \operatorname{th}(\pi r) h(r)\, dr +$$

$$+ 2 \sum_{\{P\}_\Gamma} \sum_{k=1}^{\infty} \frac{\operatorname{tr}_V \chi^k(P) \ln N(P)}{N(P)^{k/2} - N(P)^{-k/2}} g(k \ln N(P)), \tag{6.3}$$

where the sum in the right-hand side is taken over all primitive hyperbolic conjugacy classes in Γ, in the left-hand side the sum is over all solutions r_j of the equations $\frac{1}{4} + r_j^2 = \lambda_j$, where λ_j runs through the set of all eigenvalues of A.

We note that formula (6.3) gives twice the trace of the operator $K_{\Gamma,\chi}$.

We now proceed to deriving the Selberg trace formula for an arbitrary co-compact group Γ and any finite-dimensional unitary representation of it χ. Formula (6.1) is valid in this case as well, since the spectral properties of automorphic Laplacians defined both for a strictly hyperbolic group and for a more general co-compact group are, in essence, similar. The transformation (6.2) is also universal. The only difference is in the presence of a finite number of terms in the sum on the right in (6.2) which correspond to elliptic conjugacy classes in Γ. Let γ be an elliptic element in Γ; then the order of the group Γ_γ is finite. The element of Γ_γ is primitive, if it gives a clockwise rotation of the upper half-plane H through the smallest positive angle among all elements of Γ_γ.

Immediate computing gives us the assertion of the following lemma (see [163], [100]).

LEMMA 6.3. *Let $\gamma' = \gamma^k$ be a kth power of a primitive elliptic element γ of order m, ($k, m \in \mathbb{Z}$, $k \geq 1$, $m \geq 2$). Then*

$$\int_{F_{\gamma'}} k(z, \gamma' z)\, d\mu(z) = \frac{1}{2m \sin \frac{\pi k}{m}} \int_{-\infty}^{\infty} \frac{\exp[-\frac{2\pi r k}{m}]}{1 + \exp[-2\pi r]} h(r)\, dr.$$

Now, by analogy with Theorem 6.1, one can prove the following theorem. For brevity, we state it as the addition to Theorem 6.1.

THEOREM 6.2. *Suppose that Γ is an arbitrary co-compact group, χ is its unitary finite-dimensional representation, and the function h satisfies the conditions of Theorem 6.1. In this situation, the Selberg trace formula holds. It differs from (6.3) by the presence in the right-hand side of the summand*

$$\sum_{\{R\}_\Gamma} \sum_{k=1}^{m-1} \frac{\operatorname{tr}_V \chi^k(R)}{m \sin \frac{\pi k}{m}} \int_{-\infty}^{\infty} \frac{\exp[-\frac{2\pi r k}{m}]}{1 + \exp[-2\pi r]} h(r)\, dr; \tag{6.4}$$

the summation is over the set of all primitive elliptic conjugacy classes $\{R\}_\Gamma$, $m =$

$m(R)$ is the order of the class with the representative R. The rest of the summands in (6.3) are the same up to the natural dependence $r_j = r_j(\Gamma;\chi)$.

We now proceed to deriving the Selberg trace formula for the most general Fuchsian group of the first kind Γ with non-compact fundamental domain in F. It is clear that the case of the singular representation is more difficult, thus we shall consider it in detail. For a regular representation the derivation of the trace formula is in the bounds of the theory developed above, so later we limit ourselves to stating only the final result.

There are several ways of deriving the Selberg trace formula for the singular representation χ (see Chapter 13).

Below, we describe in detail one way of deriving the Selberg trace formula which, in our opinion, is the simplest way. (This is Selberg's method modified by Arthur in [39].) We also comment on some other methods.

We now consider a more simple model situation which, in fact, contains all principal difficulties of the general case. We suppose that the fundamental domain F of Γ has only one cusp and that the representation χ is trivial. For the sake of convenience, we also assume that Γ has a special reduced fundamental domain (see Chapter 4); this can always be obtained by conjugating in $G = \mathrm{PSL}(2,\mathbb{R})$, which does not change the theory essentially. Then, for simplicity, we omit in notation the dependence on χ.

We construct a special operator K_Γ for which the derivation of the Selberg trace formula is simplified. Consider the set of all operators K_Γ (see (4.5)) with the additional condition

$$k(t(z,z')) = \int_H k_1(t(z,z''))k_2(t(z'',z'))\,d\mu(z''), \tag{6.5}$$

where k_1, k_2 are certain functions of the space $C_0^\infty([0,\infty))$. We select them in such a way that the corresponding operator K_Γ is positive definite. It is not difficult to demonstrate that this choice is possible. We fix the operator K_Γ thus obtained. Now we let P_0 denote the orthogonal projection in $\mathcal{H}$ on the subspace of cusp-functions $\mathcal{H}_0$ (see Chapter 4). By the known theorem of Gelfand and Pjatetskiĭ-Shapiro the operator $P_0K_\Gamma P_0 = K_\Gamma P_0$ is compact in the space $\mathcal{H}$. More precisely, $K_\Gamma P_0$ is a Hilbert–Schmidt operator, and by (6.5) it is of trace class. One can show that the kernel of the operator in Theorem 4.3

$$T_\Gamma(z,z') = \frac{1}{4\pi}\int_{-\infty}^{\infty} h(r)E(z,\tfrac{1}{2}+ir)\overline{E(z',\tfrac{1}{2}+ir)}\,dr$$

is defined not only as a generalized function, but for the chosen operator K_Γ it is continuous and can be given by an absolutely convergent integral. Here, we make use of the positive definiteness of K_Γ. The following inequalities lie at the foundation of the proof:

$$T_\Gamma(z,z) \le k_\Gamma(z,z); \quad \left|\int_{-\infty}^{\infty} h(r)E(z,\tfrac{1}{2}+ir)\overline{E(z',\tfrac{1}{2}+ir)}\,\mathrm{d}r\right|$$
$$\le \left(\int_{-\infty}^{\infty} h(r)\left|E(z,\tfrac{1}{2}+ir)\right|^2\mathrm{d}r\right)^{1/2}\left(\int_{-\infty}^{\infty} h(r)\left|E(z',\tfrac{1}{2}+ir)\right|^2\mathrm{d}r\right)^{1/2}.$$

Since the operator $K_\Gamma P_0$ is of trace class, then from Theorem 4.3 if follows that the kernel

$$S_\Gamma(z,z') = k_\Gamma(z,z') - T_\Gamma(z,z')$$

also defines the operator of trace class in the same space. Its spectral trace is equal to $\sum_j \tilde{h}(\lambda_j)$, where λ_j runs through the set of all eigenvalues of the discrete spectrum of A. The matrix trace is equal to the following integral:

$$\int_F S_\Gamma(z,z)\,\mathrm{d}\mu(z) = \int_F \left(k_\Gamma(z,z) - T_\Gamma(z,z)\right)\mathrm{d}\mu(z). \tag{6.6}$$

Finally, the trace formula

$$\sum_j \tilde{h}(\lambda_j) = \int_F S_\Gamma(z,z)\,\mathrm{d}\mu(z) \tag{6.7}$$

is valid. To transform (6.7) to the Selberg trace formula, it is necessary to compute effectively the integral with its right-hand side. This calculation, unlike (6.1), is difficult, since, in general, the function $T_\Gamma(z,z)$ is not integrable on F in the measure $\mathrm{d}\mu(z)$. We shall do the following. Let Y be a sufficiently large positive number. We define a subdomain F^Y of F by the formula

$$F^Y = \{z \in F \mid \operatorname{Im} z = y \le Y\}.$$

We now find asymptotic expansions for the integrals

$$\int_{F^Y} T_\Gamma(z,z)\,\mathrm{d}\mu(z), \qquad \int_{F^Y} k_\Gamma(z,z)\,\mathrm{d}\mu(z).$$

The divergent principal terms in these expansions will be the same, and so they will cancel in the integral (6.6). After taking the limit as $Y \to \infty$, we will find that the other terms do give the desired value for the matrix trace on the right in (6.7). We now carry out this program in more detail. For this, we need the following special case of the Maass–Selberg relation (4.7):

$$\int_F |E^Y(z,\tfrac{1}{2}+ir)|^2\,\mathrm{d}\mu(z) = 2\ln Y - \frac{\varphi'}{\varphi}(\tfrac{1}{2}+ir)$$
$$- \frac{\overline{\varphi(\tfrac{1}{2}+ir)}Y^{2ir} - \varphi(\tfrac{1}{2}+ir)Y^{-2ir}}{2ir}. \tag{6.8}$$

We estimate the error which arises on the right in the trace formula (6.7), if we replace the function $E(z,s)$ by the function $E^Y(z,s)$ in the limit, as $Y \to \infty$. More precisely, we estimate the difference

$$\int_{F^Y} T_\Gamma(z,z)\,d\mu(z) - \int_F \frac{1}{4\pi}\int_{-\infty}^{\infty} h(r)\left|E^Y(z,\tfrac{1}{2}+ir)\right|^2 dr\,d\mu(z). \tag{6.9}$$

Here one must use the positive definiteness of the operator K_Γ. The absolute value of the difference (6.9) is bounded from above by the expression

$$\int_{F\setminus F^Y}\left(k_\Gamma(z,z) - \frac{1}{4\pi}\int_{-\infty}^{\infty} h(r)\left|y^{1/2+ir} + \varphi(\tfrac{1}{2}+ir)y^{1/2-ir}\right|^2 dr\right) d\mu(z),$$

which vanishes in the limit, as $Y \to \infty$. Taking into account (6.8), we finally get the following result:

$$\int_{F^Y} T_\Gamma(z,z)\,d\mu = g(0)\ \ln Y - \frac{1}{4\pi}\int_{-\infty}^{\infty} h(r)\frac{\varphi'}{\varphi}(\tfrac{1}{2}+ir)\,dr +$$

$$+\frac{h(0)}{4}\varphi(\tfrac{1}{2}) + O(1), \quad Y\to\infty \tag{6.10}$$

where the function g is defined in (4.4). Now it is necessary to find the asymptotic behaviour of the following integral, as $Y \to \infty$:

$$\int_{F^Y} k_\Gamma(z,z)\,d\mu(z). \tag{6.11}$$

To do this, we make use of a device which is fundamental to the derivation of the Selberg trace formula for a strictly hyperbolic group, more precisely, of the formula (6.2). The integral (6.11) is equal to the sum:

$$\sum_{\{\gamma\}_\Gamma}\ \sum_{\gamma'\in\Gamma_\gamma\setminus\Gamma}\int_{F^Y} k(t(z,\gamma'^{-1}\gamma\gamma'z))\,d\mu(z) = \sum_{\{\gamma\}_\Gamma}\int_{F^Y_\gamma} k(t(z,\gamma z))\,d\mu(z), \tag{6.12}$$

where the summation in the right-hand side of the equality is over all conjugacy classes in Γ, including parabolic ones; the domain F^Y_γ is equal by definition to $\bigcup \gamma' F^Y$, where $\gamma' \in \Gamma_\gamma\setminus\Gamma$. The domain F^Y_γ obviously becomes in the limit the fundamental domain F_γ for the centralizer Γ_γ of the element γ on the half-plane H as $Y \to \infty$. The sum (6.12) contains the terms corresponding to elliptic, hyperbolic and identity conjugacy classes. For each of them, there exists the finite limit

$$\lim_{Y\to\infty}\int_{F^Y_\gamma} k(t(z,\gamma z))\,d\mu(z) = \int_{F_\gamma} k(t(z,\gamma z))\,d\mu(z).$$

The procedure for computing this is exactly the same as in Lemmas 6.1, 6.2, 6.3 and leads to similar answers. Thus, to construct the desired asymptotic expansion of the integral (6.11), it remains for us to find the asymptotic behaviour of the sum as $Y \to \infty$

$$\sum_{\{\gamma\}_P}\int_{F^Y_\gamma} k(t(z,\gamma z))\,d\mu(z) \tag{6.13}$$

over all parabolic conjugacy classes of the group Γ. We recall that the group Γ has a reduced fundamental domain with one cusp only. It may be shown, but we shall not dwell on this question here (see [169], [39]), that, up to $o(1)$ as $Y \to \infty$, the sum (6.13) is equal to the integral

$$\int_0^Y \int_0^1 \sum_{\substack{n\in \mathbf{Z} \\ n\neq 0}} k(t(z, z+n))\, dx \frac{dy}{y^2}. \tag{6.14}$$

We shall only indicate the basic stages in computing the integral (6.14) following Selberg's lectures [163]. The sum (6.14) is equal to the following expression:

$$2\sum_{n=1}^{\infty} \tfrac{1}{n} \int_{n/Y}^{\infty} k(u^2)\, du = 2\int_0^\infty k(u^2)\Bigg(\sum_{n\leq Yu} \frac{1}{n}\Bigg)\, du = 2(\ln Y + C)\int_0^\infty k(u^2)\, du +$$

$$+ 2\int_0^\infty k(u^2) \ln u \, du + O\Big(\frac{1}{\sqrt{Y}}\Big), \quad Y \to \infty, \tag{6.15}$$

where C is Euler's constant. Along with formula (4.4), we get the following equalities:

$$2\int_0^\infty k(u^2) \ln u \, du = -\int_0^\infty \ln(1 - e^{-u})\, dg(u) + \frac{h(0)}{4} - \ln 2g(0);$$

$$-\int_0^\infty \ln(1 - e^{-u})\, dg(u) = -Cg(0) - \frac{1}{2\pi}\int_{-\infty}^{\infty} h(r)\frac{\Gamma'}{\Gamma}(1 + ir)\, dr;$$

$$\int_0^\infty k(u^2)\, du = \tfrac{1}{2} g(0),$$

where Γ is Euler's gamma-function. From this, we finally conclude that (6.13) is equal to the following sum:

$$(\ln Y - \ln 2)g(0) + \frac{h(0)}{4} - \frac{1}{2\pi}\int_{-\infty}^{\infty} h(r)\frac{\Gamma'}{\Gamma}(1 + ir)\, dr + o(1). \tag{6.16}$$

Formulas (6.7), (6.10), (6.16) and Theorem 4.3 lead to the Selberg trace formula for a model group Γ, trivial representation χ and a function h connected with a positive definite operator K_Γ. Then this formula, which we consider as the identity relative to the function h, is extended over the class of functions indicated in the assumptions of Theorem 6.1.

Before proceeding to the statement of the general theorem, we comment on other means of justifying the Selberg trace formula, i.e. the proof of decreasing the remainder term (6.9). Selberg's preliminary proof given in the lectures [163] was based upon a non-trivial *a priori* estimate for the function $E^Y(z, s)$:

$$E^Y(z, s) = O(|p(z)| \exp(3|r| - 3\,\mathrm{Im}\, z)), \quad \mathrm{Im}\, z \to \infty,$$

($s = \frac{1}{2} + ir$, $p(z)$ is a certain polynomial) and a choice of the function

$$h(r) = \widetilde{h}(\tfrac{1}{4} + r^2)$$

satisfying the condition

$$h(r) = O(\exp(-14|r|)), \quad |r| \to \infty.$$

This proof is, apparently, more difficult than that we have given above, as is justifying the Selberg trace formula from a standpoint of spectral and scattering theories (see [36], [128]).

We now give the final result (the Selberg trace formula) for an arbitrary Fuchsian group of the first kind Γ with a non-compact fundamental domain and for any unitary finite-dimensional representation χ of it (see [7]). We recall some notation (see Chapter 4) and introduce a little more: $k(\Gamma;\chi)$ is the full degree of singularity of the representation χ for the group Γ; $\{C_{\alpha l,\beta x}(s)\} = \{C_{d,b}(s)\}$ is an automorphic scattering matrix, $d = k_1 + \cdots + k_{\alpha-1} + l$, $b = k_1 + \cdots + k_{\beta-1} + k$ (this is a matrix of order $k(\Gamma;\chi)$) denoted by $C(s) = C(s;\Gamma;\chi)$; $\varphi(s) = \varphi(s;\Gamma;\chi)$ is the determinant of the matrix $C(s)$. Let P_α be the orthogonal projection in V onto the subspace V_α (see Chapter 4), and let 1_V be the identity operator in V. For any fixed $\alpha = 1,\ldots,h$ we choose a basis $e_1(\alpha),\ldots,e_n(\alpha)$ in V, in which the matrix of the operator $\chi(S_\alpha)(1_V - P_\alpha)$ is diagonal (S_α is a generator of the corresponding parabolic subgroup $\Gamma_\alpha \subset \Gamma$):

$$\chi(S_\alpha)(1_V - P_\alpha)e_l(\alpha) = \nu_{\alpha l}e_l(\alpha).$$

The following alternative is valid:

$$\nu_{\alpha l} = \begin{cases} 0, & e_l(\alpha) \in V_\alpha \\ \exp(2\pi i\theta_{\alpha l}), & e_l(\alpha) \in V \ominus V_\alpha, \end{cases}$$

where the numbers $\theta_{\alpha l}$ satisfy the inequalities $0 < \theta_{\alpha l} < 1$. One can suppose that $e_l(\alpha) \in V \ominus V_\alpha$, if $k_\alpha + 1 \le l \le \dim V$.

The theorem we state should be considered as an addition to Theorems 6.1, 6.2.

THEOREM 6.3. *For Γ and χ as above, h satisfying the conditions of Theorem 6.1, the Selberg trace formula is valid. It differs from (6.3) by the presence in the right-hand side of the additional summand (6.4) and the following summands:*

$$-2(k(\Gamma;\chi)\ln 2 + \sum_{\alpha=1}^{h}\sum_{l=k_\alpha+1}^{\dim V} \ln|1-\exp(2\pi i\theta_{\alpha l})|)g(0)+$$
$$+\frac{1}{2\pi}\int_{-\infty}^{\infty} h(r)\frac{\varphi'}{\varphi}(\tfrac{1}{2}+ir;\Gamma;\chi)\,dr + \tfrac{1}{2}(k(\Gamma;\chi)-$$
$$-\operatorname{tr} C(\tfrac{1}{2};\Gamma;\chi))h(0) - \frac{k(\Gamma;\chi)}{\pi}\int_{-\infty}^{\infty} h(r)\frac{\Gamma'}{\Gamma}(1+ir)\,dr,$$

where tr *is the trace of a matrix. The summation in the left-hand part of the trace formula obtained in such a way is over all solutions r_j of the equations $\frac{1}{4}+r_j^2 = \lambda_j$, where λ_j runs through the set of all eigenvalues of the discrete spectrum of the operator $A = A(\Gamma;\chi)$. The trace formula remains valid for a regular representation*

χ of Γ as well. In this case, one should assume the summands with the scattering matrix C and its determinant φ to be absent; in addition, $k(\Gamma;\chi)=0$.

CHAPTER 7

The Theory of the Selberg Zeta-Function

In this chapter for any Fuchsian group of the first kind we shall define the so-called Selberg zeta-function and describe its main properties. The theory we present here is interesting not only in itself, but also in that it leads to a number of spectral and geometric consequences deserving of attention. It is worth noting that all notation from the previous chapter is still valid, and we make use of it without any additional remarks.

Let Γ be an arbitrary Fuchsian group of the first kind, χ its finite-dimensional unitary representation. Following Selberg, we consider a function $Z(s;\Gamma;\chi)$ of the complex variable s, which in the region $\operatorname{Re} s > 1$ we define by means of the absolutely convergent product

$$Z(s;\Gamma;\chi) = \prod_{\{P\}_\Gamma} \prod_{k=0}^{\infty} \det(1_V - \chi(P)N(P)^{-s-k}), \tag{7.1}$$

where $\{P\}_\Gamma$ runs through the set of all primitive hyperbolic conjugacy classes in Γ, $N(P)$ is the norm of the representative P of the class $\{P\}_\Gamma$, 1_V is the identity operator in the representation space of χ. The function $Z(s;\Gamma;\chi)$ is called the Selberg zeta-function, corresponding to the group Γ and the representation χ. The reader familiar with algebraic number theory will immediately recognize some similarity in the definitions of the Selberg zeta-function and Artin's L-function. As is well recognized, in the region $\operatorname{Re} s > 1$ the L-function is given by the absolutely convergent product over all prime elements of the number field; in the Selberg zeta-function case their role is taken by the primitive hyperbolic conjugacy classes of the Fuchsian group. It will soon be clear that the Selberg zeta-function has many other important features typical of Artin's L-function.

Thus, we now proceed to clarifying the main properties of the function $Z(s;\Gamma;\chi)$. The simple calculation shows that its logarithmic derivative satisfies the following equality

$$\frac{1}{s-\frac{1}{2}} \frac{Z'(s;\Gamma;\chi)}{Z(s;\Gamma;\chi)} = \sum_{\{P\}_\Gamma} \sum_{k=1}^{\infty} \frac{\operatorname{tr}_V \chi(P^k) \ln N(P)}{N(P)^{k/2} - N(P)^{-k/2}} g(k \ln N(P); s), \tag{7.2}$$

where

$$g(u;s) = \frac{1}{2s-1} \exp(-(s-\tfrac{1}{2})|u|).$$

It is evident that the right-hand side of the formula (7.2) coincides exactly with the contribution which the hyperbolic elements of the group Γ make in the Selberg trace formula, when $h(r;s) = \left(r^2 + (s-\frac{1}{2})^2\right)^{-1}$. Unfortunately, the last function does not satisfy the conditions of Theorem 6.1, but one can consider the function $h(r;s;a) = h(r;s) - h(r;a)$ instead, where a is a complex number with $\operatorname{Re} a > 1$. If one now writes the Selberg trace formula for the function $h(r;s;a)$ corresponding to the group Γ and the representation χ, the contribution to it from the hyperbolic elements of the group will be equal to

$$\frac{1}{s-\frac{1}{2}}\frac{Z'(s;\Gamma;\chi)}{Z(s;\Gamma;\chi)} - \frac{1}{a-\frac{1}{2}}\frac{Z'(a;\Gamma;\chi)}{Z(a;\Gamma;\chi)}.$$

This special case of the Selberg trace formula is a fundamental tool for studying the zeta-function $Z(s;\Gamma;\chi)$. Thus, by calculating the integrals in this formula by means of residues, we obtain the assertion, as follows:

THEOREM 7.1. *The function $Z(s;\Gamma;\chi)$ is meromorphic on the entire complex s-plane and satisfies the functional equation*

$$Z(1-s;\Gamma;\chi) = Z(s;\Gamma;\chi)\varphi(s;\Gamma;\chi)\Psi(s;\Gamma;\chi),$$

where, by definition,

$$\Psi(s;\Gamma;\chi) = \left(\frac{\Gamma(3/2-s)}{\Gamma(s-\frac{1}{2})}\right)^{k(\Gamma;\chi)} \exp(-|F|\dim V \int_0^{s-1/2} t\,\mathrm{tg}(\pi t)\,dt +$$

$$+\pi \sum_{\{R\}_\Gamma}\sum_{k=1}^{m-1} \frac{\operatorname{tr}_V \chi(R^k)}{m\sin\frac{k\pi}{m}} \int_0^{s-1/2} \left(\frac{e^{-2\pi i k t/m}}{1+e^{-2\pi i t}} + \frac{e^{2\pi i k t/m}}{1+e^{2\pi i t}}\right) dt +$$

$$+ (1-2s)\Big(k(\Gamma;\chi)\ln 2 + \sum_{\alpha=1}^{h}\sum_{l=k_l+1}^{\dim V} \ln|1-\exp(2\pi i\theta_{\alpha l})|\Big) -$$

$$- i\arg\varphi(\tfrac{1}{2};\Gamma;\chi)),$$

and in the case of a co-compact Fuchsian group Γ or a nonsingular representation χ one must assume $k(\Gamma;\chi) = 0$, $\varphi(s;\Gamma;\chi) \equiv 1$.

In addition, if by use of the same formula one studies the residues at the poles of the logarithmic derivative of the function $Z(s;\Gamma;\chi)$, one can describe completely enough the set of zeros and poles of the Selberg zeta-function itself. The zeros of the function $Z(s;\Gamma;\chi)$ prove to be at the following points:

(1) on the line $\operatorname{Re} s = \frac{1}{2}$ symmetric relative to the real axis and on the interval $[0,1]$ symmetric relative to the point $s = \frac{1}{2}$ at the points $s_j = \frac{1}{2} + ir_j$, where $\frac{1}{4} + r_j^2 = \lambda_j$; each zero s_j has multiplicity equal to the multiplicity of the corresponding eigenvalue λ_j of the discrete spectrum of the operator $A(\Gamma;\chi)$;

(2) at the poles of the function $\varphi(s;\Gamma;\chi)$ which lie in the half-plane $\operatorname{Re} s < \frac{1}{2}$ and have the same multiplicity.

The function $Z(s;\Gamma;\chi)$ has poles at the following points

(1) $s=\frac{1}{2}$ with multiplicity $\frac{1}{2}(k(\Gamma;\chi)-\operatorname{tr}C(\frac{1}{2};\Gamma;\chi))$;

(2) $s=-l+\frac{1}{2}$, $l=1,2,\ldots$ with multiplicity $k(\Gamma;\chi)$;

(3) $s=1-\sigma_j$, where σ_j runs through the finite set of poles of the function $\varphi(s;\Gamma;\chi)$ in the interval $(\frac{1}{2},1]$ (see Chapter 4); the multiplicity of the pole at $s=1-\sigma_j$ is equal to the multiplicity of the pole σ_j of $\varphi(s;\Gamma;\chi)$.

In addition, the logarithmic derivative $Z'(s;\Gamma;\chi)/Z(s;\Gamma;\chi)$ has trivial singularities at points $s=-l$, $l=0,1,2,\ldots$, as well, the residues in which are computed by the formula

$$n_l=\frac{|F|\dim V}{\pi}(l+\tfrac{1}{2})-\sum_{\{R\}_\Gamma}\sum_{k=1}^{m-1}\frac{\operatorname{tr}_V\chi(R^k)}{m\sin\frac{k\pi}{m}}\sin\Big(\frac{k\pi(2l+1)}{m}\Big).$$

So, in the cases where $n_l=0$ the function $Z(s;\Gamma;\chi)$ has zero with multiplicity n_l at the point $s=-l$, and, when $n_l<0$ the pole with multiplicity $-n_l$. The zeta-function $Z(s;\Gamma;\chi)$ has no other zeros or poles except for those enumerated above. It should be noted that the zeros and poles are given independently of one another, and the final picture of them emerges after a comparison of the assertions stated above. So, there are no poles, given in part (3), since all of them cancel with zeros from part (1).

The results on location of zeros of the Selberg zeta-function reveal a very interesting fact: the function $Z(s;\Gamma;\chi)$ defined in terms of the group Γ and the representation χ, is highly connected with spectral characteristics of the automorphic Laplacian $A(\Gamma;\chi)$. This fact shows the apparent connection between the spectral and geometric invariants of the surface $\Gamma\backslash H$, contained in the Selberg trace formula, which will be discussed at greater length in Chapter 9.

We clarify further the properties of the Selberg zeta-function.

THEOREM 7.2. *The function $Z(s;\Gamma;\chi)$ satisfies the following relations:*

(1) $Z(s;\Gamma;\chi_1\oplus\chi_2)=Z(s;\Gamma;\chi_1)Z(s;\Gamma;\chi_2)$;

(2) *if the group Γ is a subgroup of another Fuchsian group Δ of the first kind*, then*

$$Z(s;\Gamma;\chi)=Z(s;\Delta;u^\chi), \tag{7.3}$$

where u^χ is the representation of Δ induced by the representation χ of Γ.

The first assertion of this theorem is evidently a consequence of the definition of the Selberg zeta-function. To prove the second assertion, it is sufficient to compare the logarithmic derivatives of the functions $Z(s;\Gamma;\chi)$ and $Z(s;\Delta;u^\chi)$ (the equality of them can be obtained by means of simple group-theoretic arguments; see [10]).

Note that if the group Γ is the normal divisor in Δ, and the representation χ is

---

* It is clear that the group Γ has the finite index in Δ.

trivial (i.e. $\chi \equiv 1$), then formula (7.3) is easily transformed to the form

$$Z(s;\Gamma;1) = \prod_{\psi\in(\Gamma\backslash\Delta)^*} Z(s;\Delta;\psi)^{\dim\psi}, \tag{7.4}$$

where ψ runs through the set of irreducible pairwise non-equivalent representations of the finite group $\Gamma\backslash\Delta$. One must consider formula (7.4) as a transcendental analog of Artin's factorization formula well-known in algebraic number theory. We recall that Artin's formula expresses the zeta-function of the extension of a finite degree over an algebraic number field as a product of L-functions over all irreducible representations of a corresponding Galois group. In our case, a normal subgroup Γ of finite index in Δ plays the role of an extension over a number field, and a factor group $\Gamma\backslash\Delta$ plays the role of a Galois group.

Formula (7.3) and the information on the location of the zeros of the Selberg zeta-function enable us to conclude that the discrete spectra of the automorphic Laplacians $A(\Gamma;\chi)$ and $A(\Delta;u^\chi)$ coincide exactly. Hence, by means of the Selberg trace formula, one can derive the factorization formula for the determinant of the automorphic scattering matrix $\varphi(s;\Gamma;\chi)$. This formula has the following form:

$$\Omega(\Gamma;\chi)^{1-2s}\varphi(s;\Gamma;\chi) = \Omega(\Delta;u^\chi)\varphi(s;\Delta;u^\chi), \tag{7.5}$$

where the constant $\Omega(\Gamma;\chi)$, which is connected with the group Γ and its representation χ, is defined as follows:

$$\Omega(\Gamma;\chi) = \prod_{\alpha=1}^{h} \prod_{l=k_\alpha+1}^{\dim V} |1-\exp(2\pi i\theta_{\alpha l})|.$$

It is clear that formula (7.5) makes sense only in the case where the Fuchsian group has a non-compact fundamental domain and the representation χ is singular.

The connection between the spectral characteristics of the operators $A(\Gamma;\chi)$ and $A(\Delta;u^\chi)$ discovered above has a deep intrinsic cause. The fact is that in reality, those automorphic Laplacians are unitary equivalent. Below, we provide an explicit description of the unitary isometry $T:\mathcal{H}(\Gamma;\chi)\to\mathcal{H}(\Delta;u^\chi)$ which connects the operators $A(\Gamma;\chi)$ and $A(\Delta;u^\chi)$.

Consider the partition of the group Δ on the right co-sets by Γ; let $\Delta = \bigcup_{i=1}^m \Gamma\delta_i$, where $\delta_1 = 1$, $m = [\Delta:\Gamma]$. It is well-known that the induced representation u^χ of the group Δ can be described as a unitary representation in the space V^m acting by the following formula

$$u^\chi(\delta)\Big(\sum_{i=1}^m \oplus v_i\Big) = \sum_{i=1}^m \oplus \sum_{j=1}^m \overline{\chi}(\delta_i\delta\delta_j^{-1})v_j, \quad \delta\in\Delta, \quad v_i\in V,$$

where

$$\chi(\delta) = \begin{cases} \chi(\delta), & \delta\in\Gamma, \\ 0, & \delta\notin\Gamma. \end{cases}$$

We now define the linear mapping $T : \mathcal{H}(\Gamma;\chi) \to \mathcal{H}(\Delta;u^\chi)$ by means of the formula

$$(Tf)(z) = \sum_{i=1}^{m} \oplus f(\delta_i z), \quad f \in \mathcal{H}(\Gamma;\chi), \quad z \in H.$$

The direct verification enables us to prove that this definition is correct, i.e. T maps the functions automorphic relative to the group Γ with the representation χ in the functions automorphic relative to Δ and u^χ. In addition, the mapping T is bijective, isometric and commutes with the Laplace operator on the upper half-plane H which implies $TA(\Gamma;\chi) = A(\Delta;u^\chi)T$.

We note that the factorization formulas for the Selberg zeta-function and the determinant of the automorphic scattering matrix can be proved making use of the unitary equivalence of the operators $A(\Gamma;\chi)$ and $A(\Delta;\chi)$. In fact, the following relation is valid for the kernels of the resolvents of these automorphic Laplacians:

$$\operatorname{tr} r(z,z';s;\Delta;u^\chi) = \sum_{i=1}^{m} \operatorname{tr} r(z,z';s;\delta_i^{-1}\Gamma\delta_i;\chi). \tag{7.6}$$

On the other hand, an excellent formula is presented in the paper by Faddeev [34] (see Theorem 4.2, part 2):

$$\operatorname{tr}(r(z,z';s;\Gamma;\chi) - r(z,z';1-s;\Gamma;\chi))$$
$$= \frac{1}{2s-1}\sum_{\alpha=1}^{h}\sum_{l=1}^{k_\alpha}\langle E_{\alpha l}(z;s;\Gamma;\chi), \overline{E_{\alpha l}(z';1-s;\Gamma;\chi)}\rangle_V \tag{7.7}$$

(for the vector case, this is proved in [7]). This enables us to express both parts of formula (7.6) in terms of Eisenstein series of the corresponding groups.

Now integrating as $z = z'$ the obtained equality over the reduced fundamental domain Γ^Y of the group Γ and making use of the Maass–Selberg relation, we obtain in the limit as $Y \to \infty$ the factorization formula for the determinant of the automorphic scattering matrix. The reader can provide the missing details without difficulty as on the whole, this proof of formula (7.5) is fully analogous to the corresponding part of deriving the Selberg trace formula (see Chapter 6). The formula of the factorization for the Selberg zeta-function also follows easily from here. Thus, we see that similar assertions lead to entirely different proofs – this situation is rather typical of the spectral theory of automorphic functions.

We now consider some applications of the theory of the Selberg zeta-function as developed by us here. We denote by $N(\lambda;\Gamma;\chi)$ the distribution function for the eigenvalues of the discrete spectrum of the automorphic Laplacian $A(\Gamma;\chi)$, i.e., $N(\lambda;\Gamma;\chi) = \{$ the number of eigenvalues λ_j of the operator $A(\Gamma;\chi) \mid \lambda_j \le \lambda\}$. The Selberg trace formula which is applied to the function $h(r,t) = \exp(-(\frac{1}{4}+r^2)t)$ and the Tauberian theorem enable us to obtain the following asymptotic formula

$$N(\lambda;\Gamma;\chi) = \frac{1}{4\pi}\int_{-T}^{T}\frac{\varphi'}{\varphi}(\tfrac{1}{2}+ir;\Gamma;\chi)\,dr \underset{\lambda\to\infty}{\sim} \lambda\frac{|F|\dim V}{4\pi}, \tag{7.8}$$

where $\lambda = \frac{1}{4} + T^2$, $T > 0$. We shall call it the Weyl–Selberg formula. It was first obtained by Selberg, who considered the case of scalar representation χ (i.e., $\dim V = 1$), and reported this later in his lectures [163]. In our opinion, this formula is a natural generalization of the well-known Weyl law for the main term in the asymptotic behaviour of the spectrum of the Laplace operator on a compact Riemannian manifold in the case when it does have a continuous spectrum. We note that if the group Γ is co-compact or the representation χ is non-singular then the second summand in the left-hand part of the formula (7.8) is absent and we have Weyl's ordinary law. Our task is to sharpen the remainder term in the asymptotic formula (7.8) making use of the theory of the Selberg zeta-function. More precisely, we write out explicitly two more terms of the asymptotic behaviour for the expression in the left-hand side of (7.8) (this is the spectral application of the mentioned theory).

THEOREM 7.3. *The following asymptotic formula is valid:*

$$N(\tfrac{1}{4} + T^2; \Gamma; \chi) - \frac{1}{4\pi} \int_{-T}^{T} \frac{\varphi'}{\varphi}(\tfrac{1}{2} + ir; \Gamma; \chi)\, dr$$
$$= \frac{|F| \dim V}{4\pi} T^2 - \frac{k(\Gamma; \chi)}{\pi} T \ln T + \tfrac{1}{\pi}(k(\Gamma; \chi)(1 - \ln 2) -$$
$$- \sum_{\alpha=1}^{h} \sum_{l=k_\alpha+1}^{\dim V} \ln |1 - \exp(2\pi i \theta_{\alpha l})|) T + O(T/\ln T), \quad T \to \infty. \tag{7.9}$$

As regards the proof of this theorem, it is reduced to calculating the number of zeros of the Selberg zeta-function $Z(s; \Gamma; \chi)$ on the interval $[-T, T]$ of the line $\operatorname{Re} s = \frac{1}{2}$ as $T \to \infty$ and generalizes a method well-known in analytic number theory for constructing an asymptotic formula for the number of nontrivial zeros of the Riemann zeta-function in a 'large' rectangle in the critical strip (see, for example, [175]). First of all, as can be seen from the functional equation for the function $Z(s; \Gamma; \chi)$ (see Theorem 7.1), the expression in the left-hand side of the formula (7.9) can be rewritten in the form

$$\frac{1}{2\pi} \arg \Psi(\tfrac{1}{2} + iT; \Gamma; \chi) + \frac{1}{\pi} \arg Z(\tfrac{1}{2} + iT; \Gamma; \chi) + O(1),$$

where the values of the arguments of the functions $\Psi(s; \Gamma; \chi)$ and $Z(s; \Gamma; \chi)$ are obtained by a continuous expansion from some real point $a > 1$ along the path consisting of two intervals of straight lines $[a, a + iT]$ and $[a + iT, \frac{1}{2} + iT]$. Then, by means of a simple calculation, using only the definition of the function $\Psi(s; \Gamma; \chi)$ (see Theorem 7.1), we define the asymptotic behaviour as $T \to \infty$ for the value $\arg \Psi(\frac{1}{2} + iT; \Gamma; \chi)$, which gives us the main contribution to the right-hand part of the formula (7.9). The remaining, most difficult part to prove is in establishing the estimate

$$\arg Z(\tfrac{1}{2} + iT; \Gamma; \chi) = O(T/\ln T), \quad T \to \infty.$$

This estimate is actually obtained by standard methods of analytic number theory. It turns out to be significant that the Selberg zeta-function satisfies an analog of the Riemann hypothesis 'to the right' (i.e. in the region $\operatorname{Re} s > \frac{1}{2}$) modulo a finite number of zeros at points on the interval $(\frac{1}{2}, 1]$ of the real axis. (The reader can find the details in the references for this chapter.)

We now give another application of the Selberg zeta-function. Let $\pi(\kappa; \Gamma)$ be the distribution function for the values of the norms of primitive hyperbolic conjugacy classes in Γ, i.e. $\pi(\kappa; \Gamma) = \{$ the number of primitive hyperbolic conjugacy classes $\{P\}_\Gamma \mid N(P) \leq \kappa\}$. By means of simple geometric arguments it is not difficult to prove that

$$\pi(\kappa; \Gamma) = O(\kappa), \quad \kappa \to \infty. \tag{7.10}$$

The determination of the exact order of the function $\pi(\kappa; F)$ and the refinement of the remainder term in the asymptotic formula (7.10) are based essentially on studying the properties of the Selberg zeta-function $Z(s; \Gamma; 1)$ corresponding to the trivial one-dimensional representation of the group Γ.

Let $1 = s_0 > s_1 \geq s_2 \geq \cdots \geq s_N$ denote the ordered set of all zeros of the function $Z(s; \Gamma; 1)$ in the interval $(\frac{1}{2}, 1]$ counting the multiplicity. Each such zero s_j corresponds to the eigenvalue $\lambda_j = s_j(1 - s_j) \in [0, \frac{1}{4})$ of the automorphic Laplacian $A(\Gamma; 1)$.

THEOREM 7.4. *The following asymptotic formula holds:*

$$\pi(\kappa; \Gamma) = \operatorname{li} \kappa + \sum_{j=1}^{N} \operatorname{li}(\kappa^{s_j}) + O(\kappa^{7/8+\delta}(\ln \kappa)^{-1}), \tag{7.11}$$

where li *is the integral logarithm, and* $\delta > 0$ *is an arbitrary fixed number.*

Formula (7.11) is analogous to a refined version of the prime number theorem, and is connected with the Selberg zeta-function in the same way as the asymptotic law for prime numbers is connected with the Riemann zeta-function. We note at once that a certain closed geodesic of the length $\ln N(P)$* on the factor-surface $\Gamma \backslash H$ corresponds in a one-to-one manner to any primitive hyperbolic conjugacy class $\{P\}_\Gamma$ of the Fuchsian group Γ. Thus, Theorem 7.4 can be regarded as a geometric application of the theory of the Selberg zeta-function.

The proof of Theorem 7.4 is based on a certain estimate for the logarithmic derivative of the function $Z(s; \Gamma; 1)$ and is made by means of a standard device from analytic number theory (a formula with a discontinuous Dirichlet integral). For details see [100], [7].

Summarizing the obtained results, we note that the theory of the Selberg zeta-function is analogous, in some sense, to that part of the analytical theory of the

---

* In the metric on $\Gamma \backslash H$ which is induced by the Poincaré metric on the upper half-plane H. If the group Γ contains elliptic elements, then this metric has the singularities in the images of elliptic points of Γ in $\Gamma \backslash H$.

Riemann zeta-function $\varsigma(s)$ which is based only upon the following properties of $\varsigma(s)$: representability of it in the form of an Euler product in the region $\operatorname{Re} s > 1$ and existence of a functional equation for it. On the contrary, the theory of the function $\varsigma(s)$ which is based on additive properties of its Dirichlet series has no analog for the Selberg zeta-function at present. Evidently, for the function $Z(s;\Gamma;\chi)$ there exists the Dirichlet series in the region $\operatorname{Re} s > 1$, which one can obtain after simple multiplication of the factors and combining similar terms. But we cannot describe in any efficient way the set, over which the summation is taken in the obtained series, and indicate its coefficients. So, for example, we are not able to generalize for the Selberg zeta-function the method, by means of which Hardy proved the infinity of the number of zeros of the function $\varsigma(s)$ on the line $\operatorname{Re} s = \frac{1}{2}$, since a connection of the function $\varsigma(s)$ as a Dirichlet series with a certain automorphic form is the basis of this method (this connection is carried out by the Mellin transformation). This circumstance is a matter of some difficulty in the theory of the Selberg zeta-function which makes the problem of effective representation of the function $Z(s;\Gamma;\chi)$ by a Dirichlet series rather urgent.

It should be noted that all of the assertions given above are of a general nature since they are proved for any Fuchsian group of the first kind and use none of the individual properties of certain groups. Some partial results on the location of zeros of the Selberg zeta-function will be provided in the following chapter.

CHAPTER 8

Problems in the Theory of the Discrete Spectrum of Automorphic Laplacians

This chapter is devoted to a survey of the results obtained in the theory of the discrete spectrum of the operators $A(\Gamma;\chi)$. In other words, we present the known results on zeros of the Selberg zeta-function $Z(s;\Gamma;\chi)$ lying on the half-plane $\operatorname{Re} s \geq \frac{1}{2}$ (more precisely, on the line $\operatorname{Re} s = \frac{1}{2}$ and in the interval $(\frac{1}{2}, 1]$ of the real axis, since it has no other zeros on the indicated half-plane). As before, we limit ourselves to considering only the Fuchsian groups of the first kind and their finite-dimensional unitary representations. The notation of the previous chapters is still valid.

The first questions which arise naturally when we consider the discrete spectrum of the operator $A(\Gamma;\chi)$ are the questions of the number of eigenvalues and the asymptotic behaviour of the function $N(\lambda;\Gamma;\chi)$.

In the case, when γ is co-compact or χ is non-singular, the automorphic Laplacian $A(\Gamma;\chi)$ has a purely discrete spectrum (see Chapter 4), i.e. in the Hilbert space $\mathcal{H} = \mathcal{H}(\Gamma;\chi)$ the eigenbasis for $A(\Gamma;\chi)$ exists. This means that the number of eigenvalues of the discrete spectrum for $A(\Gamma;\chi)$ is infinite. In addition, formula (7.9) is in this case as follows

$$N(\lambda;\Gamma;\chi) = \frac{|F| \dim V}{4\pi}\lambda + O(\sqrt{\lambda}), \quad \lambda \to \infty, \tag{8.1}$$

which is far stronger than simply the infinity of the number of eigenvalues for $A(\Gamma;\chi)$.

If Γ has a non-compact fundamental domain and χ is singular, then the situation is far from simple.

As we know (see Chapter 4), in this case $A(\Gamma;\chi)$ has a continuous spectrum of multiplicity $k(\Gamma;\chi)$, so the subspace $\Theta(\Gamma;\chi)$ is necessarily infinite-dimensional. But at the present time the general spectral theory cannot guarantee infinite dimensionality of the subspace of the discrete spectrum for $A(\Gamma;\chi)$. This is explained, in the first place, by the difficulties which obstruct the study of eigenvalues lying on the continuous spectrum, since there are only a finite number of eigenvalues outside the continuous spectrum of the automorphic Laplacians.

Below we familiarize the reader with what we believe to be the basic methods currently known for investigating the discrete spectrum of the operator $A(\Gamma;\chi)$, methods which enable one to prove infinite dimensionality of the subspace $\mathcal{H}_0(\Gamma;\chi)$.

It should be mentioned that, as a rule, these methods require the use of additional information, and so are primarily suitable for special groups Γ and representations χ. We shall also illustrate some difficulties which stand in the way of proving infinite dimensionality of the subspace $\mathcal{H}_0(\Gamma;\chi)$ for the general group Γ.

First of all, we shall focus our attention on the Weyl–Selberg formula (7.8). Initially, it seems possible to obtain information on the discrete spectrum of the operator $A(\Gamma;\chi)$, if we estimate the integral $\int_{-T}^{T}\frac{\varphi'}{\varphi}(\frac{1}{2}+ir;\Gamma;\chi)\,dr$ in the left-hand part of this formula. But it is difficult to get a suitable estimate for this integral in the case of general Fuchsian groups. In his lectures [163], Selberg indicated that formula (7.8) (he considered only the case of one-dimensional representation χ) will not, in general, yield any information about the asymptotic behaviour of the function $N(\lambda;\Gamma;\chi)$ as $\lambda\to\infty$, except in certain special cases of so-called arithmetic groups for which one can explicitly compute the function $\varphi(s;\Gamma;\chi)$ in terms of the Riemann zeta-function and other special functions of analytic number theory. We will now examine these cases in more detail.

We provide some standard definitions. We shall call two subgroups Γ_1 and Γ_2 of $G=\mathrm{PSL}(2,\mathbb{R})$ commensurable (in the broad sense) if there exists $g\in G$ such that the intersection $\Gamma_1\cap g^{-1}\Gamma_2 g$ has finite index in Γ_1 and in $g^{-1}\Gamma_2 g$. Let then $\Gamma_{\mathbb{Z}}$ be the modular group $\mathrm{PSL}(2,\mathbb{Z})$. It is well known that $\Gamma_{\mathbb{Z}}$ is the Fuchsian group of the first kind with signature $(0;3;2;1)$. A subgroup Γ in G commensurable with the modular group $\Gamma_{\mathbb{Z}}$ will be called an arithmetic group with a non-compact fundamental domain.

For an arbitrary arithmetic group Γ, it is very problematic to compute the function $\varphi(s;\Gamma;1)$ in terms of the special functions of analytic number theory. However, it is relatively simple to do this for the modular group $\Gamma_{\mathbb{Z}}$ and its congruence subgroups. We recall that the subgroup

$$\Gamma_1(n)=\left\{\begin{pmatrix} a & b\\ c & d\end{pmatrix}\in\Gamma_{\mathbb{Z}}\,\middle|\, a\equiv d\equiv 1(\mathrm{mod}\, n),\quad b\equiv c\equiv 0(\mathrm{mod}\, n)\right\}$$

is called the principal congruence subgroup of $\Gamma_{\mathbb{Z}}$ of level n. In the broad sense, we define a congruence subgroup as a subgroup of $\Gamma_{\mathbb{Z}}$ which contains $\Gamma_1(n)$ for some n. So-called Hecke congruence subgroups should be emphasized separately

$$\Gamma_0(n)=\left\{\begin{pmatrix} a & b\\ c & d\end{pmatrix}\in\Gamma_{\mathbb{Z}}\,\middle|\, c\equiv 0(\mathrm{mod}\, n)\right\}.$$

All of the above-mentioned groups play a significant role in the arithmetic theory of automorphic forms.

For $\Gamma_{\mathbb{Z}}$ the following formula given by Selberg in his lectures [163] is well-known:

$$\varphi(s;\Gamma_{\mathbb{Z}};1)=\sqrt{\pi}\frac{\Gamma(s-\frac{1}{2})\zeta(2s-1)}{\Gamma(s)\zeta(s)}, \tag{8.2}$$

where $\zeta(s)$ is the Riemann zeta-function, and $\Gamma(s)$ is Euler's gamma-function. The formula can be derived directly from the definition (see Chapter 4).

Next, let the natural number n be odd and squarefree. The formulas for the determinant of the automorphic scattering matrix $\varphi(s;\Gamma;1)$ in the cases where Γ coincides with one of the groups $\Gamma_0(n)$, $\Gamma_1(n)$ and also for certain other congruence subgroups were found by Hejhal. Part of his results were published in chapter 10 of [99] (see also [100]). So, for example, the following formula is valid:

$$\varphi(s;\Gamma_0(n);1) = \left(\sqrt{\pi}\frac{\Gamma(s-\frac{1}{2})\varsigma(2s-1)}{\Gamma(s)\varsigma(2s)}\right)^{h(n)} \times$$
$$\times \prod_{i=1}^{r} \frac{1}{(p_i^{2s}-1)}\left((p_i-1)^2-(p_i^s-p_i^{1-s})^2\right),$$

where $n = p_1, \ldots, p_n$ is the prime factorization, and $h(n) = 2^r$ is the number of primitive parabolic conjugacy classes of $\Gamma_0(n)$. The description of the determinant of the automorphic scattering matrix for $\Gamma_1(n)$ is more complicated, and we cannot even outline it here. Nevertheless, the results of the paper [99] imply the following theorem.

THEOREM 8.1. *Suppose that Γ is any of the groups $\Gamma_{\mathbf{Z}}$, $\Gamma_0(n)$, $\Gamma_1(n)$, where n is odd and squarefree.*

Then the distribution function for the eigenvalues of the discrete spectrum of $A(\Gamma;1)$ satisfies the Weyl asymptotic law, i.e.

$$N(\lambda;\Gamma;1) \underset{\lambda\to\infty}{\sim} \frac{|F|}{4\pi}\lambda. \tag{8.3}$$

The basis of the proof of this theorem lies with two properties of $\varphi(s;\Gamma;1)$ which hold for all of the mentioned groups Γ, namely:

(1) $\ln \varphi(s;\Gamma;1)$ is defined for $\operatorname{Re} s > 1$ by an absolutely convergent Dirichlet series;

(2) $\varphi(s;\Gamma;1)$ is a meromorphic function of order one. For details see [7].

We note that by means of the factorization formula for the determinant of the automorphic scattering matrix (7.5), it is not difficult to obtain the following result. Suppose that Γ contains the principal congruence subgroup $\Gamma_1(n)$, where n is odd and squarefree, then for the distribution function $N(\lambda;\Gamma;1)$ the Weyl formula (8.3) is valid as well. In fact, the representation u^1 of Γ induced by the trivial one-dimensional representation of its subgroup $\Gamma_1(n)$ is decomposed into the direct sum $1 \oplus (u^1 \ominus 1)$. The desired assertion follows immediately from Theorems 7.3 and 8.1 because of additivity of the function $N(\lambda;\Gamma;\chi)$ (the last means $N(\lambda;\Gamma;\chi_1\oplus\chi_2) = N(\lambda;\Gamma;\chi_1) + N(\lambda;\Gamma;\chi_2)$ which is evident).

We mention another class of arithmetic Fuchsian groups of the first kind with non-compact fundamental domain, for which the automorphic scattering matrix can be simply computed in the explicit form. This is the class of so-called cycloid subgroups of the modular group, i.e. the subgroups of $\Gamma_{\mathbf{Z}}$ which have only one cusp up to equivalence. The relevance to these subgroups of the modular group first arose because of the work of Petersson [146], [147]. For an arbitrary cycloid subgroup Z, the determinant of the automorphic scattering matrix can easily be computed by use of the following simple formula:

$$\varphi(s; Z; 1) = m^{1-2s}\varphi(s; \Gamma_Z; 1), \tag{8.4}$$

where $m = [\Gamma_Z : Z]$ is the index of Z in Γ_Z, and the function $\varphi(s; \Gamma_Z; 1)$ is given by formula (8.2). Formula (8.4) is an immediate consequence of the factorization formula (7.5) if we take into account that the representation u^1 of Γ_Z induced by the one-dimensional trivial representation of its subgroup Z, is decomposed into the direct sum of the trivial and non-singular representations. We also note that formula (8.4) can be obtained if we compare the Eisenstein series for the groups Γ_Z and Z (see Chapter 4). Thus, for any cycloid subgroup Z of the modular group, the distribution function $N(\lambda; Z; 1)$ for eigenvalues of the discrete spectrum of the automorphic Laplacian $A(Z; 1)$ has Weyl's asymptotics (8.3) as $\lambda \to \infty$.

We will now comment on the difficulties which stand in the way of the use of the Weyl–Selberg formula (7.8) for studying the asymptotic behaviour of the distribution function in the case of general Fuchsian groups of the first kind with non-compact fundamental domain. Currently, the only way of estimating the integral in the left-hand part of formula (7.7) is to a certain extent a direct computation of the function $\varphi(s; \Gamma; 1)$ to verify conditions (1) and (2) as above. Even for principal congruence subgroups of the modular group this computation is sufficiently difficult, but for general arithmetic groups it is still incomplete. For any non-arithmetic group Γ with a non-compact fundamental domain, we have no hope whatsoever of verifying the above-mentioned conditions for $\varphi(s; \Gamma; 1)$ by means of a direct computation. The reason is, that unlike the arithmetic case, the only way to define Γ is by a system of generators and relations (see Chapter 4). In practice, this method is unsuitable for computing the matrix entries for an arbitrary transformation in Γ, since it gives only the complicated recursion relations for them (see [8]), and further, those matrix entries occur in the definition of the Dirichlet series for the function $\varphi(s; \Gamma; 1)$. Even in the simplest case of the non-arithmetic triangular Hecke group (the definition of the triangular Hecke group will be given below), the constructive description of this ordered set of matrix entries is a familiar unsolved problem in the theory of Fuchsian groups*.

We now proceed to other methods of studying the discrete spectrum of the automorphic Laplacians. Suppose that the Fuchsian group Γ of the first kind is a subgroup of another Fuchsian group Δ. As indicated in the previous chapter, the operators $A(\Gamma; 1)$ and $A(\Delta; u^1)$ are unitary equivalent and, thus, they have the same eigenvalues of the discrete spectrum. The latter means that the distribution functions $N(\lambda; \Gamma; 1)$ and $N(\lambda; \Delta; u^1)$ coincide exactly, so the behaviour of the func-

---

* Here, the matrix entries can be expressed in terms of certain continued fractions (see [9], [8]), but it is as yet unclear how useful such information is for spectral theory.

tion $N(\lambda;\Delta;u^1)$ gives information on the behaviour of the function $N(\lambda;\Gamma;1)$. For example, if the representation u^1 of the group Δ contains a non-singular direct summand χ, then for all sufficiently large λ the following inequality is valid:

$$N(\lambda;\Gamma;1) > \frac{|F|\dim\chi}{4\pi[\Delta:\Gamma]}\lambda + o(\lambda),$$

which implies infinity of the set of eigenvalues of $A(\Gamma;1)$. Unfortunately, this method has a rather restricted sphere of applications since not every Fuchsian group is a subgroup of a larger Fuchsian group. Nevertheless, it enables one to prove that in any Fuchsian group of the first kind Γ there exists a subgroup Γ_0 of finite index, for which $N(\lambda;\Gamma_0;1)\xrightarrow{\lambda\to\infty}\infty$ (for details, see [7]).

We will now consider a certain special class of Fuchsian groups of the first kind, namely, the groups with non-trivial commensurators (this class also contains all subgroups of Fuchsian groups of finite index).

First, the necessary definitions. The element $g \in G = \mathrm{PSL}(2,\mathbb{R})$ lies in the commensurator $\overline{\Gamma}$ of the group Γ, if the groups Γ and $g^{-1}\Gamma g$ are commensurable in the narrow sense, i.e. the intersection $\Gamma \cap g^{-1}\Gamma g$ has finite index in Γ and in $g^{-1}\Gamma g$. It is well-recognized and not difficult to verify that the commensurator $\overline{\Gamma}$ is a group containing Γ as a subgroup. We shall say that the group Γ has a non-trivial commensurator if $\overline{\Gamma} \neq \Gamma$. Although the vast majority of the Fuchsian groups of the first kind have the trivial commensurator, a fair number of the groups Γ that arise in applications have the property that $\overline{\Gamma} \neq \Gamma$. Thus, for example, the commensurator of any arithmetic group coincides with the group $\mathrm{PSL}(2,\mathbb{Q})$ (which fully characterizes the arithmetic groups).

The Fuchsian groups with non-trivial commensurators are of particular significance from the standpoint of spectral theory. Now to explain this fact. Let g be an arbitrary element of the commensurator of the group Γ. We decompose Γ into a union of right co-sets of the subgroup $\Gamma' = \Gamma \cap g^{-1}\Gamma g$:

$$\Gamma = \bigcup_{i=1}^{m} \Gamma'\gamma_i, \quad m = [\Gamma : \Gamma'].$$

Let f be an arbitrary function on H automorphic relative to Γ (we suppose that the representation χ of the group Γ is one-dimensional and trivial). We define the operator $T(g)$ by the formula

$$(T(g)f)(z) = \sum_{i=1}^{m} f(g\gamma_i z), \quad z \in H.$$

It is evident that $T(g)$ is a linear operator taking an automorphic function to an automorphic function. In addition, as an operator in Hilbert space $\mathcal{H}(\Gamma;1)$ $T(g)$ is bounded and commutes with the automorphic Laplacian $A(\Gamma;1)$. The operators $T(g)$ are well-recognized and play an extremely important role in the arithmetic theory of automorphic forms. For the special situations these operators were known to Kronecker, and later Hurwitz. It was Hecke who studied them systematically in the theory of modular forms. We shall also call the operators $T(g)$ the (generalized) Hecke operators since this is the customary name for them in the theory of automorphic forms.

Thus, we conclude that any non-trivial (i.e., not lying in the group Γ) element $g \in \overline{\Gamma}$ generates a non-trivial Hecke operator $T(g)$ which lies in the commutator algebra of the operator $A(\Gamma; 1)$. We demonstrate that one can make use of the Hecke operators to get information on the discrete spectrum of the operator $A(\Gamma; 1)$. Suppose that the Fuchsian group Γ of the first kind has h cusps up to equivalence, and its commensurator $\overline{\Gamma}$ contains at least h^2 non-trivial elements generating h^2 different Hecke operators $T_j (j = 1, 2, \ldots, h^2)$. The Eisenstein series $E_i(z, s)$ is connected with any cusp κ_i of the group Γ (see Chapters 4, 5). For any Eisenstein series the expansion

$$T_j E_i(z, s) = \sum_{k=1}^{h} H^j_{ik}(s) E_k(z, s)$$

is valid, where the functions H^j_{ik} are defined by the formula

$$H^j_{ik}(s) = \lim_{y \to \infty} (y^{-s} T_j E_i(\sigma_k z, s)),$$

$y = \operatorname{Im} z$, σ_k are the same as in Chapter 4. This fact is easily proved by use of the arguments of the spectral theory of self-adjoint operators while studying the asymptotic behaviour of the function $T_j E_i(z, s)$ in the neighbourhood of the cusps $\kappa_1, \ldots, \kappa_h$. We denote by $H^j_l(s)$ the function $H^j_{ik}(s)$ assuming $l = (i-1)h + k$ and consider the determinant

$$\Delta(s) = \det \bigl(H^j_l(s)\bigr)^{h^2}_{j,l=1}.$$

THEOREM 8.2. *Let Γ be a Fuchsian group of the first kind with non-compact fundamental domain, and let h be a number of pairwise non-equivalent cusps. Suppose that there exist h^2 of non-trivial Hecke operators T_j which are connected with the group Γ, and the determinant $\Delta(s)$ vanishes nowhere. Then the set of eigenvalues of the discrete spectrum of the operator $A(\Gamma; 1)$ is infinite.*

We say a few words concerning the proof of this theorem. Let $k_0, k_1, \ldots, k_{h^2}$ be an arbitrary set of invariant kernels (invariants for a pair of points). Due to the results of Chapter 4, the following spectral decompositions are valid:

$$\sum_{\gamma\in\Gamma} k_0(z,\gamma z') = \sum_i \widetilde{h}_0(\lambda_i) w_i(z)\overline{w_i(z')} + \\ + \frac{1}{4\pi}\int_{-\infty}^{\infty} h_0(r) \sum_{k=1}^{h} E_k(z,s)\overline{E_k(z',s)}\,dr,$$

$$T_j\Big(\sum_{\gamma\in\Gamma} k_j(z,\gamma z')\Big) = \sum_i \Lambda_i^{(j)} \widetilde{h}_j(\lambda_j) w_i^{(j)}(z)\overline{w_i^{(j)}(z')} + \\ + \frac{1}{4\pi}\int_{-\infty}^{\infty} h_j(r) \sum_{k=1}^{h}\sum_{l=1}^{h} H_{kl}^{j}(s) E_l(z,s)\overline{E_k(z',s)}\,dr, \tag{8.5}$$

$$s = \tfrac{1}{2} + ir, \quad j = 1,2,\ldots,h^2.$$

Here we make use of the following notation: $\{w_i^{(j)}(z)\}$ is an orthonormal basis of common eigenfunctions for the operators $A = A(\Gamma;1)$ and T_j in the subspace of the discrete spectrum of the operator A. The operators A and T_j commute but the operators T_j and $T_{j'}$ when $j \neq j'$, do not, in general, commute. In addition,

$$T_j w_i^{(j)}(z) = \Lambda_i^{(j)} w_i^{(j)}(z), \quad A w_i^{(j)}(z) = \lambda_i w_i^{(j)}(z),$$
$$h_j(r) = \widetilde{h}(\tfrac{1}{4} + r^2),$$

and, finally, $\{w_i(z)\}$ is an arbitrary orthonormal eigenbasis for the operator A in the subspace $\mathcal{H} \oplus \Theta_0$. We are also supposing that the Hecke operators T_j act on the variable z. We regard the set of spectral decompositions (8.5) as a system of linear equations for the integrals with the Eisenstein series, and we regard the set of functions h_j as a set of free parameters. The condition $\Delta(s) \neq 0$ enables one to select these free parameters in such a way as to eliminate the unknown integrals adding term by term the decompositions (8.5). As a final result, we obtain the equality

$$\sum_{\gamma\in\Gamma}\Big(k_0(z,\gamma z') - \sum_{j=1}^{h^2} T_j k_j(z,\gamma z')\Big) \\ = \sum_i \Big(\widetilde{h}_0(\lambda_i) w_i(z)\overline{w_i(z')} - \sum_{j=1}^{h^2} \Lambda_i^{(j)} \widetilde{h}_j(\lambda_j) w_i^{(j)}(z)\overline{w_i^{(j)}(z')}\Big). \tag{8.6}$$

The equality (8.6) leads to a contradiction if one supposes the subspace of the discrete spectrum of A to be finite-dimensional. In fact, this supposition means that the sum in the right-hand side of (8.6) contains only a finite number of summands and, thus, this is a continuous function of the variables z, z' for any admissible kernels $k_0, k_1, \ldots, k_{h^2}$ on the left. On the other hand, one can prove that on the left in (8.6) after the special choice of the admissible kernels connected with Green's function for the differential operator $L - s(1-s)$ when $z = z'$ a logarithmic singularity arises which can only be compensated by an infinite number of summands on the right. The reader can find details of the proof in [7].

We illustrate Theorem 8.2 with a simple example of so-called Fricke groups. By definition, the Fricke group is a Fuchsian group of the first kind with signature $(1;0;1)$. These groups play an important role in Markov's arithmetic theory of the minima of quadratic forms and its generalizations (see [161]). It is well-recognized that the set of all such groups up to conjugation in $\mathrm{PSL}(2,\mathbb{R})$ forms a family which depends continuously on one complex parameter. As shown in [161], any Fricke group is a subgroup of index 2 in a Fuchsian group with signature $(0;2;2;2;1)$, i.e. it has a non-trivial commensurator. It is not difficult to see that the conditions of Theorem 8.2 hold for any Fricke group.

It should be noted that Theorem 8.2 is ineffective in the sense that, in general, it does not say anything concerning the order of the main term of the asymptotics for the distribution function $N(\lambda;\Gamma;1)$ as $\lambda \to \infty$. From this point of view, the factorization formula (7.5) gives a better result although its sphere of application is somewhat narrower. So, for the Fricke group Γ the formula (7.5) implies the estimate

$$N(\lambda;\Gamma;1) > \frac{|F|}{8\pi}\lambda + o(\lambda) = \frac{\lambda}{4} + o(\lambda)$$

for all sufficiently large values of λ (here the coefficient of λ is half as much as in Weyl's law).

Now we discuss the cases in which Theorem 8.2 allows sharpening or refinement. The broad class of such examples are Fuchsian groups commensurable with groups generated by reflections. In order to produce a detailed description we first need to extend our definition of the commensurator of the Fuchsian group. Previously, we assumed that $\overline{\Gamma} \subset G = \mathrm{PSL}(2,\mathbb{R})$. However, as will be seen later, it is useful to allow the commensurator to contain reflections relative to geodesics, i.e., motions of the second kind of the upper half-plane H. Obviously, such transformations are not contained in the group G. The condition for $\mathcal{R}$ to belong to the commensurator is the same as before: commensurability in the narrow sense of the groups Γ and $\mathcal{R}\Gamma\mathcal{R}$.

We now describe the set of Fuchsian groups of the first kind which are connected with the groups generated by reflections relative to the sides of a geodesic polygon in the Lobachevsky plane. Let M be a regular polygon in H, i.e. the polygon bounded by a finite number of geodesics, the interior angles of which are of the form π/k, where $k \in \mathbb{Z}$, $k \geq 2$ or $k = \infty$, if the corresponding vertex M belongs to the set $\mathbb{R} \cup \{\infty\}$. Let $h = h(M)$ be the number of vertices of M, lying on $\mathbb{R} \cup \{\infty\}$ (i.e. the number of zero angles in M). If $h \neq 0$, then M is non-compact; however, in all cases we consider its invariant volume $|M|$ to be finite. Then, let Γ_M^0 be the group of all motions of H, generated by reflections relative to the sides of the polygon M. We define Γ_M to be the subgroup of index two consisting of words of even length in the generators of Γ_M^0 (i.e., the subgroup of all proper motions of H). The group Γ_M is the Fuchsian group of the first kind. Its fundamental domain (in general, not canonical) has the form $F_M = M \cup \mathcal{R}M$, where $\mathcal{R}$ is a reflection relative to some side of M.

The groups Γ_M are classical groups. They were known to Klein in connection with the problem of existence of an analytic automorphic form for a given Fuchsian group. For the groups Γ_M the problem was solved in a particularly simple and elegant manner using Riemann's conformal mapping theorem and the Schwarz symmetry principle. As will be seen later, the finite dimensionality of the subspace of eigenfunctions of the discrete spectrum of $A(\Gamma_M; 1)$ can also be proved in a relatively simple manner for the groups Γ_M.

Now, a simple example of the groups Γ_M. It is a family of so-called triangular Hecke groups $\Gamma(n)$ we have mentioned above. For every natural number $n \geq 3$, we define the group $\Gamma(n)$ as a subgroup in $G = \mathrm{PSL}(2, \mathbb{R})$ generated by the transformations $z \to -z^{-1}$ and $z \to z + 2\cos(\pi/n)$ of the upper half-plane H. It is clear that the group $\Gamma(n)$ coincides up to conjugation in G with the Fuchsian group generated by reflections relative to the sides of the regular triangle with interior angles $\pi/2, \pi/n, 0$. Note that the group $\Gamma(n)$ is arithmetic only in the cases where $n = 3, 4, 6$ or ∞, and the group $\Gamma(3)$ is the modular group $\Gamma_{\mathbb{Z}}$. The Hecke groups were considered in the paper by Roelcke [157] from the standpoint of the spectral theory of automorphic functions. It was proved in this paper that for any group $\Gamma(n)$ the subspace of the discrete spectrum of $A(\Gamma(n); 1)$ is finite-dimensional. It was Roelcke who conjectured that the analogous assertion is valid in a more general situation. So, it seems appropriate to call the following assertion the Roelcke conjecture: for any Fuchsian group Γ of the first kind and for any singular representation χ the condition

$$N(\lambda; \Gamma; \chi) \xrightarrow{\lambda \to \infty} \infty$$

is fulfilled (see Chapter 13).

We now proceed to proving the Roelcke conjecture (for $\chi \equiv 1$) for the group generated by reflections relative to the sides of an arbitrary regular polygon M. First of all, we fix the reflection $\mathcal{R}$ relative to some side of M. Evidently, $\mathcal{R}$ lies in the commensurator of the group Γ_M since $\mathcal{R}\Gamma_M\mathcal{R} = \Gamma_M$ by definition. We consider a Hecke operator $T(\mathcal{R})$ connected with $\mathcal{R}$; for any automorphic function $f(z)$ $(z \in H)$

$$(T(\mathcal{R})f)(z) = f(\mathcal{R}z).$$

It is evident that the operator $T(\mathcal{R})$ is unitary, self-adjoint in the Hilbert space $\mathcal{H}(\Gamma_M; 1)$ and commutes with the automorphic Laplacian $A(\Gamma_M; 1)$. Then let $E_1(z, s), \ldots, E_h(z, s)$ denote the set of Eisenstein series connected with the group Γ_M (the group Γ_M has up to equivalence $h = h(M)$ cusps). For any of them

$$T(\mathcal{R})E_i(z, s) = E_i(z, s),$$

which one can easily prove studying the corresponding asymptotics as $\mathrm{Im}\, z \to \infty$.

Now the same argument as in Theorem 8.1 proves the validity of the Roelcke conjecture for all groups Γ_M. We note that it was sufficient here to have only one non-trivial Hecke operator.

The assertion we obtain can be considerably strengthened.

THEOREM 8.3. *Let the Fuchsian group Γ of the first kind be commensurable with some group Γ_M. Then there exists a constant c_{Γ,Γ_M} depending only on the groups Γ, Γ_M such that for all sufficiently large values of λ the inequality*

$$N(\lambda;\Gamma;1) > c_{\Gamma,\Gamma_M}\lambda$$

is valid.

If up to conjugation in $\mathrm{PSL}(2,\mathbb{R})$ *the group Γ coincides with the group Γ_M, then one can take as c_{Γ,Γ_M}*

$$\frac{|M|}{4\pi}-\epsilon \quad (\epsilon>0).$$

The proof of this theorem can be carried out making use of the spectral theory of the operator $A(\Gamma;1)$ as we have described above (see the discussion on Theorem 8.2). In addition, the specific character of the case enables one to obtain the effective estimate for the distribution function $N(\lambda;\Gamma;1)$. Details can be found in [7].

We note one remarkable consequence of Theorem 8.3. Since the modular group $\Gamma_{\mathbb{Z}}$ is a Fuchsian group generated by reflections (namely, it coincides with the Hecke group $\Gamma(3)$), one can assert that the Roelcke conjecture is valid for all the arithmetic groups.

The discussion of the questions connected with the asymptotic distribution of the eigenvalues of the discrete spectrum of automorphic Laplacians can be summed up as follows. All recognized examples of the Fuchsian groups of the first kind with non-compact fundamental domain for which the Roelcke conjecture is valid become exhausted by the groups with the non-trivial commensurators. The arithmetic groups (and what is more, only cycloidal and congruence subgroups of the modular group) give all known examples of the groups for which the Weyl law is valid. The question on existence of the exclusion of these rules is of primary interest for the spectral theory of automorphic functions (see Chapter 13).

In connection with the Fuchsian groups generated by the reflections, it is relevant to determine their role in the applications of our theory to the classical boundary value problems of mathematical physics. Let, as before, M be an arbitrary regular polygon on the upper half-plane H, let Γ_M be the corresponding Fuchsian group of the first kind, and let T be a Hecke operator connected with a reflection relative to some side of M. We consider in the Hilbert space $\mathcal{H}(\Gamma_M;1)$ the operators $P_D=\frac{1}{2}(I-T)$ and $P_N=\frac{1}{2}(I+T)$ (here I denotes the identity operator). From the properties of the operator T (see above) it follows immediately that P_D and P_N are orthogonal projectors in $\mathcal{H}(\Gamma_M;1)$ onto the mutually orthogonal subspaces $\mathcal{H}_D$

and $\mathcal{H}_N$, complementing each other (i.e. $\mathcal{H}(\Gamma_M; 1) = \mathcal{H}_D \oplus \mathcal{H}_N$). Furthermore, the restrictions of the operator $A(\Gamma_M; 1)$ to the subspaces $\mathcal{H}_D$ and $\mathcal{H}_N$ coincide respectively with the operators of the Dirichlet and Neumann boundary value problems on the polygon M for the Laplace differential operator L.

This fact enables one to use the methods of the spectral theory of automorphic functions for investigating the boundary value problems on the regular polygons.

We consider the Dirichlet problem in more detail. First of all, general methods of the spectral theory prove that the spectrum of the operator of the Dirichlet problem on any regular polygon M is purely discrete (the polygon M can be either compact or non-compact) and lies on the semi-axis $(\frac{1}{4}, \infty)$ as a whole. In addition, if $k(z, z')$ is a sufficiently good invariant kernel, then the following trace formula for the Dirichlet problem is valid:

$$\sum_j \tilde{h}(\lambda_j^D) = \frac{1}{2} \int_{F_M} \sum_{\gamma \in \Gamma_M} \left(k(z, \gamma z) - k(z, \mathcal{R}\gamma z)\right) \mathrm{d}\mu(z); \tag{8.8}$$

here λ_j^D runs through the set of all eigenvalues of the Dirichlet problem on M, $\mathcal{R}$ is the reflection relative to a fixed side of M, $F_M = M \cup \mathcal{R}M$ is the fundamental domain of the group Γ_M. As in Chapter 6, one can represent the right-hand side of (8.8) as the sum of contributions from various conjugacy classes of the group Γ_M and some relative conjugacy classes, and obtain, as a result, the Selberg trace formula for the Dirichlet boundary value problem. The reader can find the detailed derivation and the explicit form of this formula in [7].

The most interesting consequence of the modified Selberg trace formula is the theorem on the asymptotic behaviour of the distribution function $N_D(\lambda; M)$ for the eigenvalues of the Dirichlet problem on the regular polygon M.

THEOREM 8.4. *The following aymptotic formula for the Dirichlet problem is valid:*

$$N_D(\lambda; M) = \frac{|M|}{4\pi}\lambda - \frac{h}{4\pi}\sqrt{\lambda}\,\ln\lambda + \frac{1}{2\pi}(c_M + h)\sqrt{\lambda} + O\left(\frac{\sqrt{\lambda}}{\ln\lambda}\right), \quad \lambda \to \infty,$$

where h is the number of zero angles of the polygon M, c_M is a constant (see [7], formula (6.6.5)).

These asymptotics should be regarded as excellent if viewed from the standpoint of general spectral theory. It is proved by the methods outlined in Chapter 5 and is based on study of the Selberg zeta-function corresponding to the Dirichlet problem which is defined by analogy to the classical Selberg zeta-function. For further details, see [7].

As for the von Neumann problem on the regular polygon M, the results are much weaker. If the polygon is non-compact, then the spectrum of the operator of the von Neumann problem contains an absolutely continuous part of multiplicity h, and one fails to say anything of its discrete part. To all appearances, the situation here is as difficult as in the spectral theory of $A(\Gamma;1)$ for a general Fuchsian group Γ with a non-compact fundamental domain.

Now to return to studying the spectrum of automorphic Laplacians. We will investigate the eigenvalues which belong to the interval $[0,\frac{1}{4})$ (so-called small eigenvalues) or those zeros of the Selberg zeta-function which lie on the real axis on the interval $(\frac{1}{2},1]$. What is the reason for the particular interest in such eigenvalues? Firstly, as we have seen in Chapter 7, existence of small eigenvalues of $A(\Gamma;1)$ mainly affects the asymptotic behaviour of the function $\pi(\kappa;\Gamma)$, i.e. the distribution function of the norms of primitive hyperbolic conjugacy classes of a Fuchsian group Γ. A very similar situation is the problem of the number of points of a lattice in a non-Euclidean circle, and we shall discuss this at greater length. Let, as before,

$$t(z,z')=\frac{|z-z'|^2}{\operatorname{Im} z\operatorname{Im} z'}$$

be the fundamental invariant for a pair of points $z,z'\in H$ connected with the non-Euclidean distance $d(z,z')$ between z and z' by the formula

$$t(z,z')=2(\operatorname{ch} d(z,z')-1).$$

We let $N_\Gamma(T;z,z')$ denote a number of elements $\gamma\in\Gamma$ for which $t(z,\gamma z')\le T$. As $T\to\infty$, the function $N_\Gamma(T;z,z')$ has the following asymptotics (see [105], [140]):

$$N_\Gamma(T;z,z')=\frac{\pi}{|F|}T+ \\ +\sqrt{\pi}\sum_{k=1}^{N}\frac{\Gamma(s_k-\frac{1}{2})}{\Gamma(s_k+1)}w_k(z)w_k(z')T^{s_k}+O(T^{3/4}),\quad T\to\infty;$$

here, as in the previous chapters, $s_k(1-s_k)=\lambda_k$, $s_k\in(\frac{1}{2},1)$, where $\lambda_1,\ldots,\lambda_N$ are eigenvalues of $A(\Gamma;1)$ belonging to the interval $(0,\frac{1}{4})$, $w_1,\ldots,w_n$ are orthonormal eigenfunctions corresponding to them.

These statements are quite sufficient to distinguish the small eigenvalues from all other eigenvalues of automorphic Laplacians. But the interest in them is far from being exhausted by the applications of geometric nature. In representation theory, for example, it is well-acknowledged that to small eigenvalues of $A(\Gamma;1)$ there correspond the irreducible components of the special form in decomposition of the regular representation of $G=\mathrm{PSL}(2,\mathbb{R})$ in the Hilbert space $L_2(\Gamma\backslash G)$ (the so-called representations of a complementary series). However, the remarkable paper by Selberg [166], who found a deep relation between small eigenvalues and number-theoretic problems, gave rise to a keen interest in this theory. These results were discussed in part in Chapter 5 in connection with the estimates for Fourier coefficients of automorphic forms. It should be noted that the papers following [166] (see, for example, [28], [68], [17], etc.) promoted a more complete elucidation of the role the small eigenvalues play in the arithmetic applications of the spectral theory of automorphic functions.

We now proceed to discussing the known facts. We note immediately that we are only interested in the case of the trivial, one-dimensional representation of the Fuchsian group Γ, i.e. the case $\chi \equiv 1$. First of all, we comment briefly on the results which were obtained for a strictly hyperbolic Fuchsian group or, in other words, for compact Riemann surfaces of constant negative curvature equal to -1. The natural question as to whether the Laplace operator has the eigenvalues in the interval $(0, \frac{1}{4})$, or, in general, how small they can be, was set by J. Delsart [67] in 1942. In the seminal paper [164] in the footnote on page 74, Selberg indicated the possible existence in this case of a finite number of eigenvalues in the interval $[0, \frac{1}{4})$. Nevertheless, it was unclear whether such eigenvalues actually do exist, or whether the general spectral theory simply cannot guarantee their absence.

The situation became more complicated when in 1972, in a paper [132] containing some important results, (see Chapter 9) McKean proposed invalid proof of the following assertion: for any compact Riemann surface $\lambda_1 \geq \frac{1}{4}$. It was probably the stimulus for Randol who showed in [152] that in any strictly hyperbolic group Γ one can choose a normal subgroup Γ^* of finite index in such a way that the number of eigenvalues of $A(\Gamma^*; 1)$ belonging to the interval $[0, \frac{1}{4})$ will be arbitrarily large. The main technical observation of Randol (although he does not refer to this directly in [152]) is that the least eigenvalue of $A(\Gamma; \chi)$ must lie near point zero, if the representation χ differs a little from the trivial one, and his proof is based on the Selberg trace formula. As indicated in the footnote of [152], Selberg stated that he obtained this result before, together with many others, by means of the method which enables one to demonstrate directly the continuous dependence of the spectrum of $A(\Gamma; \chi)$ on the representation χ of the group Γ.

Later on, considerably more refined results were obtained (see [57], [162]). So, for example, for any number $\epsilon > 0$ and natural number $g \geq 2$, there exists the Riemann surface of genus g such that the corresponding Laplace operator has $2g-3$ eigenvalues in the interval $(0, \epsilon)$. It is in some senses impossible to strengthen this assertion, since the eigenvalue λ_{2g-2} is bounded from below by a positive constant which is the same for all the surfaces of genus g.

In addition, the total number of eigenvalues of the Laplace operator on an arbitrary surface belonging to the interval $[0, \frac{1}{4})$ does not exceed $4g-2$. This last result makes the following statement by Selberg in [164] even more precise (the note on page 74): the number of small eigenvalues of the operator $A(\Gamma; 1)$ is bounded from above by a constant proportional to the volume of the fundamental domain of the group.

Concluding the discussion of the case of strictly hyperbolic Fuchsian groups, we note that the question on existence of small eigenvalues is part of the general problem of effective estimates for eigenvalues of a Laplace operator on Riemann manifolds. This field is very extensive, and we cannot even briefly indicate here its main aspects and therefore we direct the reader to [87] and the Bibliography in it.

Before passing on to consideration of general Fuchsian groups of the first kind, we recall first that if the fundamental domain of a group is non-compact, then the eigenfunctions are divided into two types: the residues of Eisenstein series and the parabolic forms of weight zero.

This somewhat complicates the problem of studying small eigenvalues since, in addition, one must watch the behaviour of corresponding eigenfunctions. The latter means in particular, that one cannot make use of the methods of [152] in the case of general Fuchsian groups because they do not allow us to say anything on the eigenfunctions. Nevertheless, in his 1963 report [166] Selberg stated that the Eisenstein series can have the poles in the s-plane, arbitrarily close to the point $s = 1$, and described how to construct the corresponding examples in the class of arithmetic Fuchsian groups (more precisely, in the class of subgroups of finite index in the modular group). In other words, this means that the eigenvalues of automorphic Laplacians in the space of non-complete theta-series can be arbitrarily close to zero.

This example of Selberg's will be further elaborated. We denote by $\Gamma_{\mathbb{Z}}$ the modular group $\mathrm{PSL}(2,\mathbb{Z})$. The group $\Gamma_{\mathbb{Z}}$ is the Fuchsian group of the first kind with signature $(0;3;2;1)$ and is generated by the transformations $Sz = z+1$ and $Rz = -z^{-1}$ of the upper half-plane H; the first of them is parabolic, the second is an elliptic one of order 2.

The commutant Γ^* of the group $\Gamma_{\mathbb{Z}}$ is the subgroup of index 6 in $\Gamma_{\mathbb{Z}}$ with signature $(1;0;1)$. The transformations $A_1 = SRS^2$, $B_1 = S^{-1}RS^{-2}$, $S_1 = S^6$ satisfying the unique relation

$$A_1^{-1}B_1^{-1}A_1B_1S_1 = 1$$

can be given as the standard system of generators of the group Γ^*. For any element $\gamma \in \Gamma^*$ we denote by $a(\gamma)$ and $b(\gamma)$ respectively the sums of exponents with which the generators A_1 and B_1 enter the expression for γ. Then, let m and n be arbitrary mutually prime integers. We define a subgroup Γ_Q^* in Γ^* as the set of all elements $\gamma \in \Gamma^*$, for which

$$ma(\gamma) + nb(\gamma) \equiv 0 \bmod Q$$

holds, where Q is an arbitrary natural number. As Selberg indicated, for any number $\delta > 0$ and sufficiently large Q the Eisenstein series $E_\infty(z,s;\Gamma_Q^*;1)$ has the pole on the interval $(1-\delta,1)$. It is clear that Selberg's example also shows that the total number of the poles of Eisenstein series belonging to the interval $(\frac{1}{2},1]$ can be arbitrarily large.

Although the existence of the non-trivial eigenvalues in the spaces of non-complete theta-series was in fact known a comparatively long time ago, the question as to whether parabolic forms exist corresponding to small eigenvalues remained open until very recently. The positive answer to this question was given in note [17]. One can ascertain the existence of small eigenvalues in the space of parabolic forms by means of the following test for non-triviality of the spectrum of the automorphic Laplacian on the interval $(0, \frac{1}{4})$.

THEOREM 8.5. *Let Γ be a Fuchsian group of the first kind with signature $(g; m_1, \ldots, m_l; h)$. Suppose that $|F| \geq 32\pi(g+1)$ *. Then the set of eigenvalues of $A(\Gamma; 1)$ on the interval $(0, \frac{1}{4})$ is not empty and its first (non-zero) eigenvalue satisfies the inequality*

$$\lambda_1 < 8\pi(g+1)/|F|. \tag{8.7}$$

This inequality is the generalization to the case of Fuchsian groups of the first kind of the estimate for the first eigenvalue of the Laplace operator on a compact Riemann surface obtained by Yang and Yau ([185]).

We now demonstrate with the example of cycloid subgroups of a modular group that the parabolic forms can correspond to arbitrary small eigenvalues. We recall that a subgroup in $\Gamma_{\mathbb{Z}} = \mathrm{PSL}(2, \mathbb{Z})$ of finite index having only one cusp is called a cycloid subgroup. The choice of the cycloid subgroups was made because of their remarkable property that for any of them the subspace generated by residues of Eisenstein series is one-dimensional (i.e. it consists only of constant functions); this follows from formula (8.4), for example.

We now consider a decreasing by inclusion sequence of cycloid subgroups of a modular group of type $Z_{3,i}$ (the notation of Petersson from [146]). The group $Z_{3,i}$ has index $3 \cdot 2^i$ in $\Gamma_{\mathbb{Z}}$ and signature $(0; \underbrace{2, \ldots, 2}_{2^i+2}; 1)$. It is evident that $|Z_{3,i} \backslash H| = 2^i \pi$, so by Theorem 8.5 as $i \geq 5$ the spectrum of $A(Z_{3,i}; 1)$ in the interval $(0, \frac{1}{4})$ is not empty, and the estimate

$$\lambda_1^{(i)} < 2^{3-i}$$

is valid for its first eigenvalue $\lambda_1^{(i)}$, i.e., $\lambda_1^{(i)} \to 0$ as $i \to \infty$. Since for cycloid subgroups a subspace spanned on the residues at the poles of Eisenstein series consists only of constants, then $\lambda_1^{(i)}$ must correspond to a parabolic form. This example also shows that the number of small eigenvalues of the automorphic Laplacian in the space of parabolic forms may be arbitrarily large.

We now turn to estimates for the total number and multiplicities of small eigenvalues of automorphic Laplacians which are valid in the case of an arbitrary Fuchsian

---

* We recall that the volume of the fundamental domain of the Fuchsian group Γ can be computed by the Gauss–Bonnet formula:

$$|F| = 2\pi(2g - 2 + \sum_{j=1}^{l}(1 - 1/m_j) + h).$$

So, evidently, the hypothesis of the theorem is fulfilled for the groups that have a sufficiently large number of elliptic and parabolic generators in comparison with the genus.

group of the first kind. As in the case of a compact Riemann surface, the number of the eigenvalues of $A(\Gamma; 1)$ belonging to the interval $[0, \frac{1}{4}]$ is estimated from above by a constant depending only on the signature of Γ. More precisely, the interval $[0, \frac{1}{4}]$ contains no more than $4g + 2l + 2h - 2$ eigenvalues of $A(\Gamma; 1)$ (recall that g means the genus of the Fuchsian group, l and h mean the number of respectively elliptic and parabolic generators in the standard co-representation of Γ). In fact, the stronger assertion is correct (see [16]):

THEOREM 8.6. *If the fundamental domain of the Fuchsian group Γ of the first kind is a union of N disjoint geodesic triangles then the number of eigenvalues of $A(\Gamma; 1)$ on the interval $[0, \frac{1}{4}]$ does not exceed N (including the trivial eigenvalue $\lambda_0 = 0$).*

In particular, one can take the geodesic triangle with the angles 0, π/n, π/n as the fundamental domain for the Hecke triangle group. Thus we are convinced of the validity of the fact that for any Hecke group $\Gamma(n)$ the inequality $\lambda_1 > 1/4$ is fulfilled.

One can estimate not only the total number but also the multiplicities of small eigenvalues of automorphic Laplacians in terms of signature. Making use of Cheng's topology-variational arguments [63] (somewhat refined in [51]) it is not difficult to prove that the multiplicity of an eigenvalue $\lambda_k \in [0, \frac{1}{4}]$ does not exceed $4g + 2k + 1$. This result is interesting since it shows that the multiplicity of small eigenvalues of automorphic Laplacians is estimated from above by a constant which does not depend on the multiplicity of the continuous spectrum. In fact, the number of various Eisenstein series connected with the Fuchsian group Γ is arbitrarily large in comparison with its genus g, but among the residues of those Eisenstein series it turns out to be not so many linear independent eigenfunctions.

It should be noted that the estimates for a total number and multiplicities of small eigenvalues given here are far from being precise; although for small values of g they do sometimes enable one to obtain useful information. In certain special cases of arithmetic groups the stronger results of a similar sort hold. The main stimulus here is the assertion of Selberg [166] that for any congruence subgroup of a modular group the inequality $\lambda_1 \geq \frac{1}{4}$ is valid (he proved a weaker assertion $\lambda_1 \geq \frac{3}{16}$). This assertion is now called the Selberg conjecture. The confirmation or disproval of this assertion, together with an effective estimate for a multiplicity of the eigenvalue $\lambda = \frac{1}{4}$ for the congruence subgroups, would have profound consequences in number theory.

CHAPTER 9

The Spectral Moduli Problem

This short chapter is devoted to the problem which became the focus of interest due to the work of Gelfand. The problem is to what extent the spectrum of the Laplace operator on a Riemann surface defines the geometry of this surface.

So, let M be a closed Riemann surface with the Riemann metric of a constant (unique) negative curvature equal to -1. It is widely acknowledged that the surface M can be realized as a factor-surface of the upper half-plane H by the action of some strictly hyperbolic Fuchsian group Γ of the first kind. When such a realization takes place, the same Fuchsian groups up to conjugation in $\mathrm{GL}(2,\mathbb{R})$ correspond to isometric Riemann surfaces. Gelfand supposed [14] that the spectrum of $A(\Gamma;1)$ uniquely defines a strictly hyperbolic group (naturally, up to conjugation in $\mathrm{GL}(2,\mathbb{R})$). Although, as was later discovered, the Gelfand conjecture was invalid, these problems are nevertheless interesting to a wide circle of mathematicians as before, since the question of which characteristics completely define a strictly hyperbolic Fuchsian group, in other words, the question of a family of moduli for a Riemann surface, is classical and, therefore, of great theoretical importance. We proceed to a more detailed exposition of the results obtained in this field. First we give some more simple assertions.

First of all, one can note that the spectrum of $A(\Gamma;1)$ in the case of a strictly hyperbolic Fuchsian group defines its genus g and the set of numbers $\{\ln N(P)\}$ uniquely, where P runs through the set of representatives of primitive hyperbolic conjugacy classes of the group Γ*. This fact can easily be proved by use of the Selberg trace formula, if we set $h(r;t) = \cos(ar)\exp(-tr^2)$. In fact,

$$\sum_j \cos(ar_j)\exp(-tr_j^2) = \begin{cases} \dfrac{1}{2\sqrt{\pi t}}\dfrac{d(P)\ln N(P)}{N(P)^{k/2}-N(P)^{-k/2}} + \underset{t\to 0_+}{O(1)}, & a = k\ln N(P), \\ \underset{t\to 0_+}{O(1)}, & \text{for other values of } a>0. \end{cases} \tag{9.1}$$

The summation in the left-hand side of this formula is taken for all r_j, satisfying the conditions $\frac{1}{4}+r_j^2 = \lambda_j$, where λ_j runs through the whole spectrum of the operator $A(\Gamma;1)$ (in this case it is purely discrete) and $d(P)$ means the number of primitive classes $\{P\}_\Gamma$ with given value of norm $N(P)$. We know, in addition, that the genus

* The set of numbers $\{\ln N(P)\}$ is called the spectrum of lengths (of closed geodesics) of the surface $\Gamma\backslash H$ (see clarification before Theorem 7.4 on the interpretation of the numbers $\ln N(P)$).

of Γ can be computed by the formula

$$g-1=\frac{|F|}{4\pi}=\lim_{\lambda\to\infty}\frac{N(\lambda;\Gamma;1)}{\lambda} \tag{9.2}$$

which evidently follows from the Weyl asymptotics (8.1).

The formulas given above show how closely the spectral and geometric invariants of the surface $\Gamma\backslash H$ are connected. To what extent does the spectrum of the lengths of closed geodesics define the geometry of the surface? The following theorem gives a partial answer to this question.

THEOREM 9.1. *The set of mutually non-conjugate relative to* $\mathrm{GL}(2,\mathbb{R})$ *isospectral* strictly hyperbolic Fuchsian groups is at most countable.*

This theorem is almost a direct consequence of formulas (9.1) and (9.2) due to the following result.

LEMMA 9.1. *Let Γ be a strictly hyperbolic Fuchsian group of the first kind, and let $\gamma_1,\dots,\gamma_{2g}$ be a standard system of its generators. Then the set of all numbers of the form*

$$\mathrm{Tr}\,\gamma_i,\quad \mathrm{Tr}\,\gamma_i\gamma_j,\quad \mathrm{Tr}\,\gamma_i\gamma_j\gamma_k, \tag{9.3}$$

where the indices i,j,k satisfy the conditions $1\le i\le 2g$, $i<j\le 2g$, $1<k\le 2g$, define uniquely the group Γ up to conjugacy in $\mathrm{GL}(2,\mathbb{R})$ *or, similarly, up to conjugacy in* $\mathrm{PSL}(2,\mathbb{R})$ *and the reflection*

$$\begin{pmatrix} a & b\\ c & d\end{pmatrix}\longrightarrow\begin{pmatrix} a & -b\\ -c & d\end{pmatrix}.$$

To prove this lemma, we introduce the notation

$$\gamma_i=\begin{pmatrix} a_i & b_i\\ c_i & d_i\end{pmatrix},\qquad 1\le i\le 2g.$$

Let the groups Γ and Γ' have the coincident sets of the numbers (9.3). One may suppose that the equality $\gamma_1=\gamma_1'=\begin{pmatrix}\mu & 0\\ 0 & \mu^{-1}\end{pmatrix}$, $\mu>1$, holds, which can always be obtained by means of the conjugacy from $\mathrm{PSL}(2,\mathbb{R})$.

Then the following equalities

$$\mathrm{Tr}\,\gamma_i=a_i+d_i=\mathrm{Tr}\,\gamma_i'=a_i'+d_i',$$
$$\mathrm{Tr}\,\gamma_1\gamma_i=\mu a_i+\mu^{-1}d_i=\mathrm{Tr}\,\gamma_1'\gamma_i'=\mu a_i'+\mu^{-1}d_i'$$

hold, which implies $a_i=a_i'$, $d_i=d_i'$, and also $b_ic_i=b_i'c_i'$ for all $i>1$. Then the products b_ic_j and $b_i'c_j'$ are also equal for $i\ne j$ which can be seen from the relations

$$\mathrm{Tr}\,\gamma_i\gamma_j-\mathrm{Tr}\,\gamma_i'\gamma_j'=b_ic_j-b_i'c_j'+c_ib_j-c_i'b_j'=0,$$
$$\mathrm{Tr}\,\gamma_1\gamma_i\gamma_j-\mathrm{Tr}\,\gamma_1'\gamma_i'\gamma_j'=\mu(b_ic_j-b_i'c_j')+\mu^{-1}(c_ib_j-c_i'b_j')=0,\quad 1<i<j.$$

* i.e., having the same spectrum of automorphic Laplacians connected with them.

It is necessary for us to find an element $g \in \mathrm{GL}(2,\mathbb{R})$ such that the equality $\gamma_i = g\gamma_i' g^{-1}$ holds for all values of i. We shall try to find this transformation g in the group $\mathrm{PSL}(2,\mathbb{R})$. From the equality $\gamma_1 = \gamma_1'$ it follows that g must have the form $\begin{pmatrix} \lambda & 0 \\ 0 & \lambda^{-1} \end{pmatrix}$, $\lambda > 0$. One can choose an appropriate number $\lambda > 0$ if the quotients

$$\frac{b_i}{b_i'} = \frac{c_i'}{c_i}, \quad i > 1, \tag{9.4}$$

do not depend on the choice of i and are positive. The first condition is evidently fulfilled (the numbers b_i, b_i', c_i, c_i' cannot vanish, otherwise we obtain a contradiction with the discreteness of the groups Γ and Γ'). As for the second condition, we can fulfill this if we consider the group $\begin{pmatrix} 1 & 0 \\ 0 & -1 \end{pmatrix} \Gamma' \begin{pmatrix} 1 & 0 \\ 0 & -1 \end{pmatrix}$ instead of Γ', i.e. the group conjugate to Γ' by means of the reflection

$$\begin{pmatrix} a & b \\ c & d \end{pmatrix} \longrightarrow \begin{pmatrix} a & -b \\ -c & d \end{pmatrix}.$$

Now we can clarify Theorem 9.1. In fact, the set of numbers $|\mathrm{Tr}\,\gamma| = N(\gamma)^{1/2} + N(\gamma)^{-1/2} = 2\,\mathrm{ch}(\ln N(\gamma)/2)$, $\gamma \in \Gamma$, is defined uniquely by the spectrum of $A(\Gamma; 1)$ (see formula (9.1)). The finite set of numbers of the form (9.3) up to the choice of signs must belong to the set mentioned above.

We now give a more refined result. In the context of the previous theorem it may be stated as follows:

THEOREM 9.2. *Most finite numbers of pairwise, non-conjugate strictly hyperbolic, Fuchsian groups of the first kind can have the same spectrum.*

What is the idea behind proving this theorem? One can always choose the generators $\gamma_1, \ldots, \gamma_{2g}$ of a strictly hyperbolic Fuchsian group of genus g in such a way that they would pairwise identify those sides of corresponding fundamental domains which lie through each other (see Chapter 4). By such a choice of generators, the numbers $|\mathrm{Tr}\,\gamma_i|$, $|\mathrm{Tr}\,\gamma_i\gamma_j|$ and $|\mathrm{Tr}\,\gamma_i\gamma_j\gamma_k|$ are bounded from above by the values 2, 4 or 6 respectively multiplied by $\mathrm{ch}(\mathrm{diam}\,\Gamma\backslash H)$, where $\mathrm{diam}\,\Gamma\backslash H$ means as usual the maximum of the distances between two points on the surface $\Gamma\backslash H$. We let $l_0 = \min_{\gamma\in\Gamma,\gamma\neq 1} \ln N(\gamma)$; in other words, l_0 is the length of the shortest geodesic on the surface $\Gamma\backslash H$. We note that the inequality

$$\tfrac{1}{2} l_0 \,\mathrm{diam}\,\Gamma\backslash H \leq |\Gamma\backslash H| = 4\pi(g-1) \tag{9.5}$$

is always valid. This implies that the numbers (9.3) do not exceed the values 2, 4 or 6 respectively multiplied by $\mathrm{ch}(8\pi(g-1)/l_0)$. We now consider the ordered sequence

$$2 < \cdots \leq |\mathrm{Tr}\,\gamma| \leq \cdots, \quad \gamma \in \Gamma, \quad \gamma \neq 1.$$

Evidently, this sequence has no accumulation points and so the set of numbers (9.3) can be selected in it only by a finite number of ways.

The inequality (9.5) we have used here has a simple geometric explanation. Let us take on the surface $\Gamma\backslash H$ the segment of the geodesic which realizes $\operatorname{diam}\Gamma\backslash H$, and at each point of it we take an orthogonal segment of a geodesic of the length $l_0/4$ to each of the sides. The segments which go out at the same point cannot meet, since l_0 is the length of the shortest homotopically non-trivial closed path on $\Gamma\backslash H$. The segments which go out of the different points are also disjointed by virtue of specific properties of geometry of the surfaces of negative curvature. Thus, the surface $\Gamma\backslash H$ contains an injectively imbedded strip of the volume $\geq l_0/2 \operatorname{diam}\Gamma\backslash H$, which implies the desired inequality immediately.

Further progress made in the spectral moduli problem can be characterized as follows. From the spectrum of $A(\Gamma; 1)$ one can uniquely restore (up to conjugation from $\mathrm{GL}(2,\mathbb{R})$) a strictly hyperbolic group Γ 'almost always'. To clarify this result, it is necessary to introduce the notion of a Teichmüller space.

We consider the set of pairs of the form $(M; [f])$, where M is a closed Riemann surface of genus $g \geq 2$, $[f]$ is a homotopic class of an orientation-preserving homeomorphism f of a fixed surface M_0 onto M. The pairs $(M_1; [f_1])$ and $(M_2; [f_2])$ will be called equivalent, if there exists a conformal mapping of M_1 onto M_2 homotopic to $f_2 \cdot f_1^{-1}$. The classes of equivalent pairs $(M; [f])$ will be called marked Riemann surfaces. On the set T_g of all marked Riemann surfaces of genus g one can induce naturally a real-analytic structure of a Euclidean space $\mathbb{R}^{6g-6}$. The set T_g provided with such a structure will be called a Teichmüller space.

The Teichmüller space can also be described in the language of Fuchsian groups. Consider the set of pairs of the form $(\Gamma; \jmath)$, where Γ is a strictly hyperbolic Fuchsian group of genus g, $\jmath$ is an isomorphism of the fundamental group $\pi_1(M_0)$ of the surface M_0 onto the group Γ.

We call the pairs $(\Gamma_1; \jmath_1)$ and $(\Gamma_2; \jmath_2)$ equivalent if there exists a transformation $g \in \mathrm{GL}(2,\mathbb{R})$ such that $g\Gamma_1 g^{-1} = \Gamma_2$ and $g\jmath_1 g^{-1} = \jmath_2$. The classes of equivalent pairs $(\Gamma; \jmath)$ or, in other words, the marked Fuchsian groups correspond one-to-one to the points of the space T_g, i.e., to the marked Riemann surfaces.

THEOREM 9.3. *In the Teichmüller space T_g one can determine a proper, real analytic submanifold V_g so that any strictly hyperbolic Fuchsian group Γ realizing the point in $T_g\backslash V_g$ will be defined uniquely by the spectrum of $A(\Gamma; 1)$ up to conjugation from* $\mathrm{GL}(2,\mathbb{R})$.

Finally, we give another result which will probably be interesting for the specialists in number theory. This theorem gives a sufficient condition for a group Γ to belong to the set $T_g\backslash V_g$. We introduce a notation $\operatorname{Tr}\operatorname{sp}(\Gamma) = \{\operatorname{Tr}\gamma\}_{\gamma\in\Gamma}$.

THEOREM 9.4. *Let Γ be a strictly hyperbolic Fuchsian group of genus g. Suppose that the degree of transcendence of the field $\mathbb{Q}(\operatorname{Tr}\operatorname{sp}(\Gamma))$ over $\mathbb{Q}$ is equal to $6g-6$.*

Then the group Γ *is uniquely defined (up to conjugation from* $\mathrm{GL}(2,\mathbb{R})$*) by the spectrum of* $A(\Gamma;1)$.

We do not try to prove here Theorems 9.3 and 9.4, since this would involve digression from the subject. Further information on the spectral theory of moduli can be found in Chapter 13.

CHAPTER 10

Automorphic Functions and the Kummer Problem

In this chapter we acquaint the reader with the solution of the familiar number-theoretic problem of Kummer on the distribution of the arguments of Gauss cubic sums at primes of a corresponding field. This solution was obtained comparatively recently by Heath-Brown and Patterson, and here certain ideas and methods of the spectral theory of automorphic functions previously developed by Kubota play an essential role. Those ideas will be obvious to the reader familiar with chapters 4 and 5 of this book.

First, to recall some number-theoretic definitions. We define a symbol of quadratic residue or a Legendre symbol by the following formula:

$$\left(\frac{p}{q}\right) = \begin{cases} 1, & \text{if the congruence } x^2 \equiv p(\operatorname{mod} q) \text{ is solvable,} \\ -1, & \text{otherwise;} \end{cases} \tag{10.1}$$

here p and q are primes (not necessarily positive) or ± 1, and $p \neq q$. Gauss proved that for odd primes p and q, $p \neq q$, the following formula is valid:

$$\left(\frac{p}{q}\right)\left(\frac{q}{p}\right) = (-1)^{(p-1)/2 \cdot (q-1)/2} \tag{10.2}$$

The equality (10.2) is called the quadratic reciprocity law.

We define the Gauss sum for N, $k \in \mathbb{Z}_+$ and a Dirichlet character $\chi \operatorname{mod} N$ of degree k by the following formula

$$S_k(N;\chi) = \sum_{n(\operatorname{mod} N)} \chi(n) e\left(\frac{n}{N}\right). \tag{10.3}$$

Gauss found that for an odd prime p the equality

$$S_2\left(p;\left(\frac{\cdot}{p}\right)\right) = \sqrt{p} \cdot \begin{cases} 1, & \text{if } \ p \equiv 1(\operatorname{mod} 4), \\ i, & \text{if } \ p \equiv 3(\operatorname{mod} 4), \end{cases} \tag{10.4}$$

is valid. Formula (10.4) is the basis of one proof of the reciprocity law (10.2) belonging to Gauss.

The main difficulty in proving (10.4) was in finding the value of argument of the Gauss sum, since the value of its module (as the value of the module of the general Gauss sum) is easily determined from the definition. The similar explicit general formulas for arguments of Gauss sums of a higher power are unknown at present, in spite of the titanic efforts of many outstanding mathematicians to find them in view of the great importance of explicit formulas for number theory.

At the same time, with the search for the explicit formulas, the natural question arose on distribution of the arguments of general Gauss sums. We now discuss the first difficult case of the cubic Gauss sum.

Consider the field of algebraic numbers $\mathbb{Q}(\omega)$, where $\omega = e(\frac{1}{3})$ (see List of notations). We denote by $\mathcal{O}$ the ring of integers of the field $\mathbb{Q}(\omega)$. $\mathcal{O}$ is the ring with unique factorization; all of its ideals are principal; it has 6 units, these are the numbers ϵ with $\epsilon^6 = 1$. Each number $h \in \mathcal{O}$, $h \neq 0$ can be represented uniquely in the form $\epsilon\lambda^l h'$, ϵ is the unit, $l \in \mathbb{Z}_+$, $h' \in \mathcal{O}$, $h' \equiv 1 (\mathrm{mod}\, 3)$, $\lambda = \sqrt{-3}$.

Then we introduce the function

$$g(c) = \sum_d \left(\frac{d}{c}\right)_3 \mathbf{e}\left(\frac{d}{c}\right), \tag{10.5}$$

or, more generally,

$$g(r,c) = \sum_d \left(\frac{3d}{c}\right)_3 \mathbf{e}\left(\frac{3rd}{c}\right), \quad (g(1,c) = g(c)); \tag{10.6}$$

here $c \in \mathcal{O}$, $c \equiv 1 (\mathrm{mod}\, 3)$, $\mathbf{e}(z) = \exp(2\pi i(z + \bar{z}))$ for $z \in \mathbb{C}$, $r \in \lambda^{-3}\mathcal{O}$, $\lambda = \sqrt{-3}$; the summation is taken over $d \in \mathcal{O}$, $d (\mathrm{mod}\, c)$, $(d,c) = 1$*; $(-)_3$ is a residue symbol of degree 3. We give here the definition of a residue symbol. If $\pi \in \mathcal{O}$ is a prime and $(\pi, 3) = 1$, then $q = N(\pi) \equiv 1 (\mathrm{mod}\, 3)$; here $N : \mathbb{Q}(\omega) \to \mathbb{Q}$ is a norm of a number in the field. If $a \in \mathcal{O}$, $(a, \pi) = 1$ then $a^{q-1} \equiv 1 (\mathrm{mod}\, \pi)$ is the Fermat little theorem. So, there exists (and only one) $\xi \in \mathcal{O}$, $\xi^3 = 1$, for which $a^{(q-1)/3} \equiv \xi (\mathrm{mod}\, \pi)$. We let by definition

$$\left(\frac{a}{\pi}\right)_3 = \xi.$$

Then, if $c \in \mathcal{O}$, $(c, 3) = 1$ and $c = \pi_1^{k_1} \cdot \pi_2^{k_2} \cdots \pi_l^{k_l} \cdot \epsilon$ is a canonical prime factorization (ϵ is a unit), and if $a \in \mathcal{O}$, $(a,c) = 1$, then we assume

$$\left(\frac{a}{c}\right)_3 = \prod_{r=1}^{l} \left(\frac{a}{\pi_r}\right)^{k_r}.$$

We now introduce the Gauss cubic sum. If $\pi \in \mathcal{O}$ is a prime $\equiv 1 (\mathrm{mod}\, 3)$, then $p = N(\pi)$ is a prime in $\mathbb{Z}$ and the function $g(\pi)$ is equal to

$$g(\pi) = \sum_{\substack{n (\mathrm{mod}\, p) \\ n \in \mathbb{Z}, (n,p)=1}} \left(\frac{n}{\pi}\right)_3 \mathbf{e}\left(\frac{n}{p}\right),$$

---

* We denote by (d,c) the greatest common divisor of the numbers d, c; we mean by this the number $\equiv 1 (\mathrm{mod}\, 3)$ generating the ideal $d\mathcal{O} + c\mathcal{O}$ (but not the ideal itself, as it should be done in accordance with general algebraic number theory).

i.e. the following equality is valid:

$$g(\pi) = S_3\Big(p; \Big(\frac{\cdot}{\pi}\Big)_3\Big),$$

where S_3 is the Gauss sum.

We now find the module of the Gauss sum. Let $\widetilde{g}(c) = g(c)N(c)^{-1/2}$.

It is easy to demonstrate that

$$\widetilde{g}(c)^3 = \mu(c)\cdot c|c|^{-1},$$

where μ is the Möbius function ($\mu(c) = 0$, if c is not squarefree; $\mu(c) = (-1)^l$, if $c = \pi_1\cdots\pi_l\cdot\epsilon$, where ϵ is a unit; π_j are distinct primes of $\mathcal{O}$). In particular, for prime $\pi\in\mathcal{O}$ we obtain the formula

$$\widetilde{g}(\pi)^3 = -\pi|\pi|^{-1}.$$

This property defines $g(\pi)$ up to multiplication by a cube root of the unit.

What is this root for a given number π? This was the question that Kummer examined (see [118]). After an unsuccessful attempt to find explicit formulas, he divided all primes π into three classes I, II, and III in accordance with the alternative

$$\mathrm{Re}(\widetilde{g}(\pi)) < -\tfrac{1}{2},\quad -\tfrac{1}{2}\le \mathrm{Re}(\widetilde{g}(\pi)) < \tfrac{1}{2},\quad \tfrac{1}{2}\le \mathrm{Re}(\widetilde{g}(\pi)).$$

Let for j =I, II, III, $N_j(X)$ is the number of the π of the class j such that $N(\pi)\le X$. The calculations showed (for not very large X) that $N_{\mathrm{II}}(X)$ is in twice, and $N_{\mathrm{III}}(X)$ is in three times more than $N_{\mathrm{I}}(X)$. But later calculations carried out for a larger X, suggested an idea that together with the growth of X, N_{I}, N_{II}, N_{III} increase with the same speed. This latter assertion was proved by Patterson and Heath-Brown and those were the final results of the solution to the Kummer problem.

These were their results: let Λ be the Mangoldt function in $\mathcal{O}$ and li be the integral logarithm, i.e.:

$$\Lambda(c) = \begin{cases} \log N(\pi), & \text{if } c=\pi^k\epsilon,\quad \pi \text{ is a prime} \in\mathcal{O},\ \epsilon \text{ is the unit, } k\in\mathbb{Z}_+\\ 0, & \text{otherwise.}\end{cases}$$

THEOREM 10.1. (I) *For all $l\in\mathbb{Z}$ and all $X\in\mathbb{R}$, $X\ge 2$, the estimate*

$$H_l(X) = \sum_{\substack{c\in\mathcal{O},c\equiv1(\mathrm{mod}\,3)\\ N(c)\le X}} \widetilde{g}(c)\Lambda(c)\Big(\frac{\bar c}{|c|}\Big)^l \ll X^{30/31+\epsilon} + |l|X^{29/31+\epsilon} \tag{10.7}$$

holds for any $\epsilon>0$; a constant in the symbol $\ll$ depends only on ϵ.

(II) *There exists $A\in\mathbb{R}_+$ such that for $0\le\theta_1\le\theta_2\le2\pi$, $X\ge2$, $\theta_1,\theta_2,X\in\mathbb{R}$ the following equality holds*

$$\sum_{\substack{N(\pi)\le X\\ \theta_1\le\arg\widetilde{g}(\pi)\le\theta_2}} 1 = \frac{\theta_2-\theta_1}{2\pi}\,\mathrm{li}\,X + O(X\exp(-A\sqrt{\log X})); \tag{10.8}$$

here the summation is taken over the primes $\pi \in \mathcal{O}$, $\pi \equiv 1 (\mathrm{mod}\, 3)$, satisfying the conditions $N(\pi) \le X$, $\theta_1 \le \arg \tilde{g}(\pi) \le \theta_2$; a constant in the symbol O depends only on A.

As is known from the theorem on primes,

$$\sum_{N(\pi)\le X} 1 = \mathrm{li}\, X + O(X \exp(-A\sqrt{\log X})).$$

So (10.8) means 'the uniformity of distribution of $\arg \tilde{g}(\pi)$ on a circle' which gives the solution to the Kummer problem. We note that the result (10.7) is stronger than (10.8).

We now outline the proof of the theorem trying to throw light on the role of the spectral theory of automorphic functions.

For $l \in \mathbb{Z}$ and a squarefree $\alpha \in \mathcal{O}$, $\alpha \equiv 1 (\mathrm{mod}\, 3)$ we set*

$$\psi_{(\alpha)}(s, r, l) = \sum_{\substack{c\in \mathcal{O}, c\equiv 1 (\mathrm{mod}\, 3)\\ c\equiv 0 (\mathrm{mod}\, \alpha)}} g(r, c)\left(\frac{c}{|c|}\right)^l N(c)^{-s}; \qquad (10.9)$$

$s \in \mathbb{C}$, the series converges absolutely, if $\mathrm{Re}\, s > \frac{3}{2}$. For a specialist in analytic number theory, it is quite natural to consider the Dirichlet series (10.9) in order to prove (10.7), (10.8). The ideas of Patterson and Heath-Brown may be summarized as follows. Let the functions (10.9) have been continued meromorphically to the left, and let information about them necessary for the method of complex integration have been obtained (see, for example, [171]). Then by use of this method, the estimate for the summator function of the coefficients of the series (10.9) can be obtained, i.e., for sums

$$\sum_{\substack{c\in \mathcal{O}, c\equiv 1 (\mathrm{mod}\, 3)\\ c\equiv 0 (\mathrm{mod}\, \alpha), N(c)\le X}} \tilde{g}(r, c)\left(\frac{c}{|c|}\right)^l, \quad X \to \infty. \qquad (10.10)$$

Having obtained the estimates for the sums (10.10), one can make use of the arguments of the method of trigonometric sums of Vinogradov [13] in order to obtain the estimate for the sum

$$\sum_{\substack{c\in \mathcal{O}\\ c\equiv 1 (\mathrm{mod}\, 3), N(c)\le X}} \tilde{g}(r, c)\Lambda(c)\left(\frac{c}{|c|}\right)^l, \quad X \to \infty. \qquad (10.11)$$

($= H_l(X)$, if $r = 1$). More precisely, Heath-Brown and Patterson model the method of estimating a linear trigonometric sum over primes

$$\sum_{\substack{1\le c\le X\\ c\in \mathbb{Z}}} e(\alpha c)\Lambda(c), \quad X \to \infty.$$

---

* We may take $r = 1$ at once, but there is no necessity in this, since one can study the series with an arbitrary r in the same way as with $r = 1$.

Use of the Vaughan identity turns out to be very effective [176]. Thus we obtain the estimate (10.7). One can arive at the equality (10.8) from (10.7) by means of the Kubilyus estimates [21] for the sums

$$\sum_{\substack{\pi \text{ is prime } \in \mathcal{O} \\ \pi\equiv 1 (\mathrm{mod}\, 3), N(\pi)\leq X}} \left(\frac{\pi}{|\pi|}\right)^l, \quad X\to\infty,$$

and the Weyl criterion of uniform distribution in the form of the Erdös–Turan theorem [77].

In reality, many difficulties arise in this way, and it requires skilful use of the methods of analytic number theory to overcome them. But it is not necessary to delve deep in this area, and we return to the investigation of the functions $\psi_{(\alpha)}(\cdot, r, l)$. The method was outlined by Kubota [113], [115]. To describe it, we postpone consideration of the problem and direct our attention to Fourier expansions for Eisenstein series (see Chapter 5).

Since the Eisenstein series allow a meromorphic continuation, the Fourier coefficients in the expansion (5.13) and, therefore, the Dirichlet series φ_{mk} in (5.14) can also be continued meromorphically. Next, the coefficients in the expansion (5.13) evidently satisfy the same functional equation as Eisenstein series themselves.

It is clear from the Maass–Selberg relation that if a point is a pole of some function φ_{mk}, then it is the pole of the Eisenstein series and, therefore, it is the pole of the constant term of the Eisenstein series (see Chapters 4 and 5). Finding the poles of the constant terms of the Eisenstein series is not usually a difficult problem. Thus, the theory of the Eisenstein series provides a tool for investigating a large class of Dirichlet series which appear as coefficients in Fourier expansions.

We now return to functions (10.9). Kubota found that these functions can also be studied by the method we indicated above, but the Eisenstein series with necessary properties are defined not on the homogeneous space $H = \mathbb{R}\times\mathbb{R}_+ \cong \mathrm{SL}_2(\mathbb{R})/\mathrm{SO}(2)$ but on the homogeneous space $H' = \mathbb{C}\times\mathbb{R}_+ \cong \mathrm{SL}_2(\mathbb{C})/\mathrm{SU}(2)$ (and corresponding discrete group is contained in $\mathrm{SL}_2(\mathbb{C})$, but not in $\mathrm{SL}_2(\mathbb{R})$). Here, we deviate from the theory to which the main part of our book is devoted. It should be said that replacing H by H' gives rise only to technical difficulties.

The Maass–Selberg relation is also valid in this case, and this is the main point.

Now to investigate the functions (10.9) in more detail. Let $\mathbb{F}$ be a totally imaginary field of algebraic numbers containing all the unit roots of degree n for a certain n. Let $\mathcal{O}_\mathbb{F}$ be the ring of integers of the field $\mathbb{F}$, μ_n be the group of unit roots of degree n, $(-)_n$ be the symbol of residue of degree n for $\mathbb{F}$, q be an ideal of the ring $\mathcal{O}_\mathbb{F}$ and Γ_q be the principal congruence subgroup mod q in $\mathrm{SL}_2(\mathcal{O}_\mathbb{F})$, i.e.,

$$\Gamma_q = \left\{\begin{pmatrix} a & b \\ c & d\end{pmatrix} \in \mathrm{SL}_2(\mathcal{O}_\mathbb{F}) \mid a, d \in 1+q; b, c\in q\right\}.$$

T. Kubota proved the following remarkable theorem.

THEOREM 10.2. *We define the map*

$$\chi : \Gamma_q \to \mu_n$$

by the equality

$$\chi\left(\begin{pmatrix} a & b \\ c & d \end{pmatrix}\right) = \begin{cases} \left(\frac{c}{a}\right)_n, & \text{if } c \neq 0, \\ 1, & \text{if } c = 0. \end{cases}$$

If $q \equiv 0(\text{mod}\, n^2)$ *(in some cases the condition* $q \equiv 0(\text{mod}\, n)$ *is sufficient), then* χ *is a homomorphism of groups.*

The proof of this theorem is rather simple, but it makes use of the reciprocity law. Not wishing to go into detail, we note only that for $\mathbb{F} = \mathbb{Q}(\omega)$ one can take $n = 3$, $q = (3)$. We now give certain algebra-geometrical information necessary for the definition of the Eisenstein series in the Fourier coefficients of which the Dirichlet series with the Gauss cubic sums enter.

Each matrix $\gamma \in \mathrm{SL}_2(\mathbb{C})$ can be uniquely represented by the product

$$\begin{pmatrix} 1 & z \\ 0 & 1 \end{pmatrix}\begin{pmatrix} \sqrt{v} & 0 \\ 0 & \frac{1}{\sqrt{v}} \end{pmatrix}\xi,$$

$z \in \mathbb{C}$, $v \in \mathbb{R}_+$, $\xi \in \mathrm{SU}(2)$; it is a partial case of the Iwasawa decomposition. It is easy to verify that the map $\gamma \cdot \mathrm{SU}(2) \to (z, v)$ is an isomorphism of real differentiable manifolds $\mathrm{SL}_2(\mathbb{C})/\mathrm{SU}(2)$ and $H' = \mathbb{C} \times \mathbb{R}_+$. The natural action (the multiplication from the left) of the group $\mathrm{SL}_2(\mathbb{C})$ on $\mathrm{SL}_2(\mathbb{C})/\mathrm{SU}(2)$ induces the action of this group on H'. If we represent the point $w = (z, v)$ by the matrix

$$w = \begin{pmatrix} z & -v \\ v & \bar{z} \end{pmatrix}$$

and if we denote the matrix

$$\begin{pmatrix} p & 0 \\ 0 & \bar{p} \end{pmatrix}$$

by $\tilde{p}$ for $p \in \mathbb{C}$, then the action of the element $\gamma \in \mathrm{SL}_2(\mathbb{C})$ on $w \in H'$ may be written in the form

$$\gamma w = (\tilde{a}w + \tilde{b})(\tilde{c}w + \tilde{d})^{-1}, \quad \gamma = \begin{pmatrix} a & b \\ c & d \end{pmatrix}. \tag{10.12}$$

For $w \in H'$ we define $z(w) \in \mathbb{C}$, $v(w) \in \mathbb{R}_+$ by the condition: $w = (z(w), v(w))$. The equality (10.12) is equivalent to the following two equalities:

$$\begin{aligned} z(\gamma w) &= \frac{(az+b)\overline{(cz+d)} + a\bar{c}v^2}{|cz+d|^2 + |c|^2 v^2}, \\ v(\gamma w) &= \frac{v}{|cz+d|^2 + |c|^2 v^2}. \end{aligned} \tag{10.13}$$

An invariant Riemann metric, an invariant measure and an invariant differential operator are defined on H'. The explicit formulas are not presented here since they are irrelevant to what follows.

For $\gamma \in \mathrm{SL}_2(\mathbb{C})$ and $w \in H'$ we define a function

$$j(\gamma, w) = (\tilde{c}w + \tilde{d}) \det(\tilde{c}w + \tilde{d})^{-1/2},$$

it is the 2×2 matrix. For any matrix M we denote by $\overline{M}$ a complex conjugate matrix, by M^t a transposed matrix, $M^* = \overline{M}^t$. If $l \in \mathbb{Z}_+$ then M_l is lth tensor power of the matrix M, $M_0 = 1$, and, evidently, $(MN)_l = M_l N_l$ for any $M, N \in \mathrm{SL}_2(\mathbb{C})$. It is clear that $(M^*)_l = (M_l)^*$; we denote both sides of the equality by M_l^*. It is not difficult to verify the equalities

$$\begin{aligned} j(g, hw) &= j(gh, w), \\ j(g, w)^* &= j(g, w)^{-1}, \end{aligned} \tag{10.14}$$

which is valid for all $w \in H$, $g, h \in \mathrm{SL}_2(\mathbb{C})$.

The fundamental domain of the group Γ_3 in H has 12 cusps; it may be chosen in such a way that the points ∞, 0, ± 1, $\pm\omega$, $\pm\omega^2$, $\pm(1-\omega)$, $\pm(1-\omega)^{-1}$ will be the vertices; we denote this domain by $\mathcal{D}$. The representation χ (the Kubota homomorphism) is singular only in four vertices from all of them. They are ∞, 0, ± 1; we denote the set of these points by P_0. We recall the definition: χ is singular in a cusp p, if $\chi(\gamma) = 1$ for all $\gamma \in \Gamma_{3,p} = \{\gamma \in \Gamma_3 \mid \gamma(p) = p\}$. Let then

$$\sigma_\infty = \begin{pmatrix} 1 & 0 \\ 0 & 1 \end{pmatrix}, \quad \sigma_0 = \begin{pmatrix} 0 & -1 \\ 1 & 0 \end{pmatrix}, \quad \sigma_1 = \begin{pmatrix} 1 & 0 \\ 1 & 1 \end{pmatrix}, \quad \sigma_2 = \begin{pmatrix} 1 & 0 \\ -1 & 1 \end{pmatrix}.$$

Evidently, $\sigma_p \infty = p$.

After Kubota, we can now define Eisenstein series as follows:

$$E_l(w, p, s) = \sum_{g \in \Gamma_{3,p} \backslash \Gamma_3} \overline{\chi}(g) j(\sigma_p^{-1} g, w)_l^* v(\sigma_p^{-1} g w)^s; \tag{10.15}$$

here, χ is the Kubota homomorphism, $p \in P_0$; every summand is a $2^l \times 2^l$ matrix. The series converge absolutely and uniformly on any compact set in the domain $\operatorname{Re} s > 2$, so their sums are regular functions in this domain. Then, using (10.14) one can easily verify the equality

$$E_l(\gamma w, p, s) = \chi(\gamma) j(\gamma, w)_l E_l(w, p, s)$$

for all $\gamma \in \Gamma_3$, so E_l are automorphic functions with rather exotic systems of multipliers. (This is much simpler in the case where $l = 0$. Then the multiplier system is the Kubota homomorphism.)

Examining the series (10.15) leads to the following theorem on the functions $\psi_{(\alpha)}$.

THEOREM 10.3. *Let the numbers* α, r, l *be from formulas* (10.6), (10.9) *and* $(\alpha, r) = 1$.

(1) *The function* $s \to \psi_{(\alpha)}(s,r,l)$ *can be continued meromorphically onto the entire s-plane. If* $l \neq 0$, *then it is regular in the region* $\operatorname{Re} s > 1$, *and if* $l = 0$, *it is regular at all points of the mentioned region except possibly for the point* $4/3$, *where it may have a simple pole.*

(2) *Let r be a squarefree number such that* $r \equiv 1 (\operatorname{mod} 3)$. *Then the following equality is valid:*

$$\operatorname*{Res}_{s=4/3} \psi_{(\alpha)}(s,r,0) = c_0 \frac{\overline{g(1,r)}}{N(\alpha)N(r)^{2/3}} \prod_{\substack{\pi \text{ is a prime } \in \mathcal{O} \\ \pi/\alpha, \pi \equiv 1 (\operatorname{mod} 3)}} \left(1 + N(\pi)^{-1}\right)^{-1}, \quad (10.16)$$

where

$$c_0 = (2\pi)^{5/3} 3^{-3/2} 8^{-1} \Gamma\left(\tfrac{2}{3}\right)^{-1} \zeta_{\mathbb{Q}(\omega)}(2)^{-1}, \quad (10.17)$$

$\zeta_{\mathbb{Q}(\omega)}$ *is the Dedekind zeta-function of the field* $\mathbb{Q}(\omega)$.

(3) *Let* $\epsilon > 0$ *and* $\sigma_1 = 3/2 + \epsilon$. *If* $s = \sigma + it$, $\sigma_1 \geq \sigma \geq \sigma_1 - \frac{1}{2}$, $|s - 4/3| \geq 1/12$, *then the following estimate is valid:*

$$\psi_{(\alpha)}(s,r,l) \underset{\epsilon}{\ll} N(\alpha)^{3/2(\sigma_1-\sigma)-1} N(r)^{1/2(\sigma_1-\sigma)} \left(1 + l^2 + t^2\right)^{\sigma_1 - \sigma}. \quad (10.18)$$

This theorem supplies all necessary information on the functions $\psi_{(\alpha)}$. More precisely, in order to solve the Kummer problem, it is unnecessary to know the explicit formula for the residue (10.16). It is sufficient to have the estimate

$$\operatorname*{Res}_{s=4/3} \psi_{(\alpha)}(s,1,0) \underset{\epsilon}{\ll} N(\alpha)^{-3/4+\epsilon}$$

which can be obtained from the Phragmén–Lindelöf principle without using the very difficult Theorem 10.1 [141].

We now turn to the proof of Theorem 10.3. First of all, it can be seen from (10.15) that the Eisenstein series are periodic functions

$$E_l(w + \nu, p, s) = E_l(w, p, s),$$

if $\nu \in \mathcal{O}$, $\nu = 0 (\operatorname{mod} 3)$. Hence we find the Fourier expansion

$$E_l((z,v),p,s) = \delta_{p,\infty} I_l v^s + \sum_{\nu \in \lambda^{-3}\mathcal{O}} v^{2-s} K_{s,l}(\nu v) \psi_p^+(s,\nu,l)\, e(\nu z) \quad (10.19)$$

by means of calculations analogous to those in Chapter 5. Here $\lambda = \sqrt{-3}$ so that $\lambda^{-3}\mathcal{O}$ is a lattice dual relative to $\mathbf{e}$ to $3\mathcal{O}$; I_l is the unit $2^l \times 2^l$ matrix; in addition,

$$K_{s,l} = \int_{\mathbb{C}} \left(|z|^2 + 1\right)^{-s-l/2} \begin{pmatrix} z & -1 \\ 1 & \bar{z} \end{pmatrix}_l^* e(-\alpha z)\, dz, \quad (10.20)$$

$$\psi_p^+(s,\nu,l) = V^{-1} \sum_{g \in \Gamma_{3,p} \backslash \Gamma_3 / \Gamma_{3,\infty}} \chi(g)^{-1} e\left(\frac{\nu d(g)}{c(g)}\right) \left(c(\bar{g})\right)_l^* |c(g)|^{-2s-l}, \quad (10.21)$$

$V = 9\sqrt{3}/2$ is the volume of the fundamental domain of the group Γ_3 in the invariant measure. In (10.21) $c(g)$ and $d(g)$ are defined by g from the equality

$$\sigma_p^{-1} g = \begin{pmatrix} * & * \\ c(g) & d(g) \end{pmatrix}.$$

(Summation in (10.21) is taken for all representatives of double conjugacy classes of $\Gamma_{3,p}\backslash\Gamma_3/\Gamma_{3,\infty}$. One representative is taken in each class; the class for representative g of which $c(g) = 0$ is omitted). We note that $\psi_p^+(s,\nu,l)$ is a diagonal matrix.

Then we have to continue the Eisenstein series (10.15) meromorphically onto the entire s-plane and to derive functional equations for them.

This can be done by means of the Maass–Selberg relation, and some notation is introduced here to describe it. Suppose that for $p, q \in P_0$

$$r(p,q) = \begin{cases} \infty, & \text{if} \quad p = q; \\ 0, & \text{if} \quad p,q \text{ are } 0,\infty;\ \infty,0;\ 1,0;\ -1,0; \\ 1, & \text{if} \quad p,q \text{ are } -1,\infty;\ \infty,1;\ 0,-1;\ 1,-1; \\ -1, & \text{if} \quad p,q \text{ are } 1,\infty;\ 0,1;\ -1,1;\ \infty,-1; \end{cases}$$

$$\epsilon(p,q) = \begin{cases} -1, & \text{if} \quad p,q \text{ are } 0,\infty;\ 0,-1; \\ 1, & \text{for all other pairs } p,q. \end{cases}$$

For $q \in P_0$ suppose that $y_q \in \mathbb{R}$, $y_q \geq 1$ and $C_q = \{w \in \mathcal{D} \mid v(\sigma_q^{-1}w) > y_q\}$ (pairwise disjoint neighbourhoods of cusps). We set

$$E_l^Y(w,p,s) = j_l(\sigma_q, \sigma_q^{-1}w)\epsilon(p,q)^l \{E_l(\sigma_q^{-1}w, r(q,p), s) -$$
$$- \delta_{r(q,p),\infty} I_l v(\sigma_q^{-1}w)^s - v(\sigma_q^{-1}w)^{2-s} K_{s,l}(0)\psi_{r(q,p)}^+(s,0,l)\},$$

if $w \in C_q$ for some q;

$$E_l^Y(w,p,s) = E_l(w,p,s),$$

if $w \in \mathcal{D}\backslash \bigcup_{q\in P_0} C_q$; here $Y = (y_q)_{q\in P_0}$. The Maass–Selberg relation has the form

$$(t(2-t) - s(2-s)) \int_{\mathcal{D}} E_l^Y(w,p,\bar{t})^* E_l^Y(w,q,s) \frac{dz\, dv}{v^3}$$
$$= V \sum_{\rho\in P_0} (\epsilon(\rho,p)\epsilon(\rho,q))^l \Big\{ \delta_{p,\rho}\delta_{q,\rho}(s-t) y_\rho^{s+t-2} I_l +$$
$$+ (2-s-t)\delta_{p,\rho} y_\rho^{t-s} K_{s,l}(0)\psi_{r(\rho,q)}^+(s,0,l) +$$
$$+ (s+t-2)\delta_{\rho,q} y_\rho^{s-t} \psi_{r(\rho,p)}^+(\bar{t},0,l)^* K_{\bar{t},l}(0)^* +$$
$$+ (t-s)\psi_{r(\rho,p)}^+(\bar{t},0,l)^* K_{\bar{t},l}(0)^* K_{s,l}(0)\psi_{r(\rho,q)}^+(s,0,l) y_\rho^{2-s-t} \Big\},$$

$s, t \in \mathbb{C}$, $\mathrm{Re}\, s$, $\mathrm{Re}\, t > 2$. From this relation it follows that the Eisenstein series can be continued meromorphically onto the entire s-plane and satisfy the following functional equations:

$$E_l(w,q,s) = \sum_{p\in P_0} E_l(w,p,2-s) K_{s,l}(0)\psi_{r(p,q)}^+(s,0,l)\epsilon(p,q)^l, \tag{10.22}$$

$q \in P_0$. In addition, only those points which are the poles of a certain function $s \to K_{s,l}(0)\psi^+_{r(q,p)}(s,0,l)$, $q \in P_0$ are the poles of the function (10.15) and, thus, of the function (10.21).

Consequently, the functions $s \to \psi^+_q(s,\nu,l)$ can be extended meromorphically onto the entire s-plane and satisfy the functional equations:

$$K_{s,l}(\nu v)\psi^+_q(s,\nu,l) = \sum_{p \in P_0} v^{2s-2}K_{2-s,l}(\nu v)\psi^+_p(2-s,\nu,l)K_{s,l}(0)\psi^+_{r(p,q)}(s,0,l)\epsilon(p,q)^l. \quad (10.23)$$

We recall that $\psi^+_q(s,\nu,l)$ is a diagonal $2^l \times 2^l$ matrix. If $l > 0$, then $\psi_q(s,\nu,l)$ is its first diagonal element (i.e., the element in the intersection of the first line and the first column); if $l < 0$, we take the last diagonal element of $\psi^+_q(s,\nu,l)$ as $\psi_q(s,\nu,l)$ (i.e., the element in the intersection of the last line and the last column); if $l = 0$, we set $\psi_q(s,\nu,l) = \psi^+_q(s,\nu,l)$. The computations show that for all $l \in \mathbb{Z}$ the following formulas are valid

$$\psi_\infty(s,\nu,l) = V^{-1} \sum_{\substack{c \in \mathcal{O}, c \neq 0 \\ c \equiv 0 (\mathrm{mod}\, 3)}} \sum_{\substack{d \equiv 1 (\mathrm{mod}\, 3) \\ d (\mathrm{mod}\, 3c)}} \left(\frac{c}{d}\right)_3 \mathrm{e}\left(\frac{\nu d}{c}\right)\left(\frac{\bar{c}}{|c|}\right)^l |c|^{-2s} \quad (10.24)$$

and for $q \neq \infty$

$$\psi_q(s,\nu,l) = V^{-1} \sum_{\substack{c \in \mathcal{O} \\ c \equiv \gamma(p) (\mathrm{mod}\, 3)}} \sum_{\substack{d (\mathrm{mod}\, 3c) \\ d \equiv \delta(p) (\mathrm{mod}\, 3)}} \left(\frac{d}{c}\right)_3 \mathrm{e}\left(\frac{\nu d}{c}\right)\left(\frac{\bar{c}}{|c|}\right)^l |c|^{-2s}, \quad (10.25)$$

where $\gamma(p)$ and $\delta(p)$ are defined from the equality

$$\sigma_p^{-1} = \begin{pmatrix} * & * \\ \gamma(p) & \delta(p) \end{pmatrix}.$$

Let, for the sake of brevity,

$$\psi = \psi_{(1)}. \quad (10.26)$$

The functions defined in (10.24) and (10.25) are connected with ψ by the following simple relations

$$\begin{aligned} \psi_0(s,\nu,l) &= V^{-1}(-1)^l\psi(s,\nu,l), \\ \psi_1(s,\nu,l) &= V^{-1}(-1)^l \mathrm{e}(-\nu)\psi(s,\nu,l), \\ \psi_{-1}(s,\nu,l) &= V^{-1}\mathrm{e}(\nu)\psi(s,\nu,l), \\ \psi_\infty(s,\nu,l) &= \sum_{b=2}^{\infty} \sum_{\substack{\epsilon \in \mathcal{O} \\ \epsilon^6 = 1}} \Gamma(\nu,\epsilon\lambda^b)|\epsilon\lambda^b|^{-2s}\left(\frac{\overline{\epsilon\lambda}^b}{|\epsilon\lambda^b|}\right)^l \psi(s,\epsilon\lambda^b\nu,l). \end{aligned} \quad (10.27)$$

In the last equality, we use the notation

$$\Gamma(\nu,\epsilon\lambda^b)=\sum_{\substack{\delta(\operatorname{mod}\lambda^{b+2})\\ \delta\in\mathcal{O},\delta\equiv1(\operatorname{mod}3)}}\left(\frac{\epsilon\lambda^b}{\delta}\right)_3\mathbf{e}\left(\frac{\nu\delta}{\epsilon\lambda^b}\right).$$

It is simple to verify that the set of those ϵ, b for which $\Gamma(\nu,\epsilon\lambda^b)\neq0$ is finite for any ν. The equalities (10.27) are proved by an immediate calculation. Meromorphic continuation of the function $s\to\psi(s,\nu,l)$ follows at once from any of the first three equalities (10.27), since the functions $s\to\psi_q^+(s,\nu,l)$ have been extended already.

Next, it is easy to prove the following equalities.

$$\begin{aligned}
\psi_0(s,0,l)&=V^{-1}(-1)^l\varsigma(3s-3,l)\varsigma(3s-2,l)^{-1},\\
\psi_1(s,0,l)&=V^{-1}(-1)^l\varsigma(3s-3,l)\varsigma(3s-2,l)^{-1},\\
\psi_{-1}(s,0,l)&=V^{-1}\varsigma(3s-3,l)\varsigma(3s-2,l)^{-1},\\
\psi_\infty(s,0,l)&=V^{-1}(1+(-1)^l)(3^{3s-3}i^l-1)^{-1}\varsigma(3s-3,l)\varsigma(3s-2,l)^{-1}.
\end{aligned}$$

Here we make use of the notation

$$\varsigma(s,l)=\sum_{\substack{c\in\mathcal{O}\\ c\equiv1(\operatorname{mod}3)}}\left(\frac{\bar{c}}{|c|}\right)^{3l}|c|^{-2s};\tag{10.29}$$

$\varsigma(s,l)$ is the function known in the theory of L-series of Hecke.

It is easy to find from (10.20) that for $\alpha\neq0$ the element lying in the intersection of the first line and the first column of the matrix $K_{s,l}(\alpha)$ is equal to

$$i^{-l}(2\pi)^{s+l/2}\left(\frac{\alpha}{|\alpha|}\right)^l|\alpha|^{s+(l/2)-1}\Gamma(s+\tfrac{l}{2})^{-1}K_{s-(l/2)-1}(4\pi|\alpha|),$$

and the element lying in the intersection of the last line and the last column of the matrix $K_{s,l}(\alpha)$ is equal to

$$i^{-l}(2\pi)^{s+l/2}\left(\frac{\bar{\alpha}}{|\alpha|}\right)^l|\alpha|^{s+(l/2)-1}\Gamma(s+\tfrac{l}{2})^{-1}K_{s-(l/2)-1}(4\pi|\alpha|).$$

The analogous computation for $\alpha=0$ shows that in the first line of the matrix $K_{s,l}(0)$ only one element lying in the intersection of the first line and the last column is not 0, and it is equal to

$$\pi(s+\tfrac{l}{2}-1)^{-1},$$

and in the first column of the matrix $K_{s,l}(0)$ only the element of the first column and the last line is different from 0 and is equal to

$$\pi(-1)^l(s+\tfrac{l}{2}-1)^{-1},$$

Due to this property, one can reduce the functional equations (10.23) to the functional equations for (10.24), (10.25). From these functional equations and formula

(10.27) one can derive the functional equation for the function ψ:

$$F(s,\nu,l) = A(s,l)\cdot 3^{9(1-s)}|\nu|^{2(1-s)}\left(\frac{\overline{\nu}}{|\nu|}\right)^l \times \Big\{F_\infty(2-s,\nu,-l)+$$
$$+ F(2-s,\nu,-l)\Big(\mathrm{e}(\nu)+(-1)^l\mathrm{e}(-\nu)+\frac{1+(-1)^l}{3^{3s-3}i^l-1}\Big)\Big\}; \quad (10.30)$$

in this equation we make use of the following notation

$$F(s,\nu,l) = (2\pi)^{-2s}\Gamma(s+\tfrac{|l|}{2}-\tfrac{1}{3})\Gamma(s+\tfrac{|l|}{2}-\tfrac{2}{3})\psi(s,\nu,l)\varsigma(3s-2,l),$$
$$A(s,l) = \begin{cases} 3^{-1}i^{-|l|}(1+3^{3-3s}i^l)(1-3^{3s-4}i^l)^{-1}, & \text{if } l\equiv 0 (\mathrm{mod}\, 2),\\ (\sqrt{-3})^{-1}3^{3-3s}, & \text{if } l\not\equiv 0 (\mathrm{mod}\, 2),\end{cases}$$
$$F_\infty(s,\nu,l) = \sum_{b=2}^{\infty}\sum_{\substack{\epsilon\in O\\ \epsilon^6=1}}\Gamma(\nu,\epsilon\lambda^b)3^{-bs}(\epsilon i^b)^{-l}F(s,\epsilon\lambda^b\nu,l).$$

As indicated, only the points which are the poles of any of the functions $s\to K_{s,l}(0)\psi^+_{r(q,p)}(s,0,l)$, $q\in P_0$ are the poles of the function (10.21). The more refined arguments based on the Maass–Selberg relation show that only the points which are the poles of any of the functions $s\to\psi_{r(q,p)}(s,0,l)$, $q\in P_0$, may be the poles of the functions $s\to\psi_p(s,\nu,l)$. Furthermore, we find from (10.28) that the functions $s\to\psi_p(s,\nu,l)$ have no poles in the half-plane $\operatorname{Re} s>1$ except for the point $s=4/3$ (this pole is simple) and only for $l=0$ (since the function $s\to\varsigma(s,l)$ for $l\neq 0$ is regular in the half-plane $\operatorname{Re} s\geq 1$, and for $l=0$ it has only one pole in this half-plane at the point $s=1$ and this pole is simple). The estimate (10.18) for $\psi=\psi_{(1)}$ is derived from the functional equation (10.30) making use of the Phragmén–Lindelöf principle.

So we obtain all the assertions of Theorem 10.3 about $\psi=\psi_{(1)}$, except for the explicit formula for the residue, but we shall not explain the derivation of this formula, since it is not necessary for solving the Kummer problem (see the remark after the statement of the theorem). All the assertions of the theorem on $\psi_{(\alpha)}$ can be obtained from the corresponding assertions on ψ by means of the following formula

$$\psi_{(\alpha)}(s,r,l) = \Delta_\alpha^{-1}(s,l)\sum_{\substack{d\in O, d|\alpha\\ d\equiv 1(\mathrm{mod}\,3)}}\mu(d)N(d^2)\Omega(d^2)\Omega(\alpha)g(r,\tfrac{\alpha}{d})\psi(s,\tfrac{r\alpha}{d},l); \quad (10.31)$$

Here we denote

$$\Delta_\alpha(s,l) = \prod_{\substack{\pi \text{ is a prime } \in O\\ \pi|\alpha,\, \pi\equiv 1(\mathrm{mod}\,3)}}\Big(1-N(\pi)^{2-3s}\Big(\frac{\pi}{|\pi|}\Big)^{3l}\Big),\quad \Omega(c)=N(c)^{-s}\Big(\frac{\overline{c}}{|c|}\Big)^l.$$

Formula (10.31) evidently follows from two formulas:

$$\widetilde{\psi}_{(\alpha)}(s, \alpha r, l) = \Delta_\alpha^{-1}(s, l) \sum_{\substack{d \in \mathcal{O}, d \mid \alpha \\ d \equiv 1 (\mathrm{mod}\, 3)}} \mu(d) N(d) \Omega(d^2) \overline{g(\tfrac{r\alpha}{d}, d)} \psi(s, \tfrac{r\alpha}{d}, l), \tag{10.32}$$

$$\psi_{(\alpha)}(s, r, l) = \Omega(\alpha) g(r, \alpha) \widetilde{\psi}_{(\alpha)}(s, r\alpha, l); \tag{10.33}$$

here

$$\widetilde{\psi}_{(\alpha)}(s, r, l) = \sum_{\substack{c \in \mathcal{O}, c \equiv 1(3) \\ (c, \alpha) = 1}} g(r, c) N(c)^{-s} \left(\frac{\bar{c}}{|c|}\right)^l.$$

Formulas (10.32), (10.33) can easily be verified, which completes the proof of Theorem 10.3.

CHAPTER 11

The Selberg Trace Formula on the Reductive Lie Groups

In Chapters 4 and 6 we outlined the fundamental ideas and methods of deriving the Selberg trace formula on the Lobachevsky plane. This chapter is devoted to the most general situation when the derivation of analogous formula has a certain sense. We begin with the classical statement of the problem according to Selberg (see [165], [164]).

Let X be a Riemann space, and let G be a locally compact group of isometries of X. The group G acts on X, and, hence, G acts on every function f defined on X

$$f(x) \to f(gx). \tag{11.1}$$

For definiteness we suppose $f : X \to \mathbb{C}$. In (11.1) $x \in X$, $g \in G$. We consider the ring V of linear integro-differential operators on X permutable with the action (11.1) for any $g \in G$. We suppose in the following that V is finitely generated and commutative (the assumption lays a strong constraint on X and G). We pick out a basis $\mathcal{D}_1, \dots, \mathcal{D}_n$ in V. We shall denote it sometimes by a single letter $\mathcal{D}$. It turns out to be a sufficiently general case when one can pick out as $\mathcal{D}_1, \dots, \mathcal{D}_n$ symmetric differential operators. For any operator $K \in V$ there exists a function h such that

$$K = h(\mathcal{D}_1, \dots, \mathcal{D}_n) \equiv h(\mathcal{D}). \tag{11.2}$$

On the other hand, the integral operator K is defined by the kernel $k : X \times X \to \mathbb{C}$. Permutability of K with the action (11.1) means

$$k(gx, gx') = k(x, x'). \tag{11.3}$$

So, we have the map of the functions $k \to h$ which turns out to be invertible. If X is a symmetric space (for definition, see below), then this map is the Harish-Chandra transformation for general spherical functions which is the analog of the Fourier transformation.

Now, let $\Gamma \subset G$ be a discrete group of transformations of X such that there exists a compact fundamental domain F with $F \cong \Gamma \backslash X$. Each operator of the ring V can be considered as the operator of the Hilbert space $L_2(F; d\mu)$, where the measure $d\mu$ is generated by the metric of X (invariant relative to G, of course). By virtue of compactness of F, each $\mathcal{D}_j$ as the operator in $L_2(F; d\mu)$ will be the operator with a discrete spectrum. Thus, there exists an orthonormal system of its eigenfunctions f_i which is full in $L_2(F; d\mu)$, such that

$$\mathcal{D}_j f_i(x) = \lambda^i_j f_i(x),$$

where the sequence $\lambda^i = (\lambda^i_1, \ldots, \lambda^i_n)$ has no accumulation points in $\mathbb{R}^n$.

The operator $K \in V$ with the kernel $k(x, x')$ considered as the operator in $L_2(F; d\mu)$ (we let K_Γ denote this operator) is defined by the kernel

$$k_\Gamma(x, x') = \sum_{\gamma \in \Gamma} k(x, \gamma x'). \tag{11.4}$$

We suppose that $k(x, x')$ has the decrease necessary for the series (11.4) to be absolutely convergent. Then, for a sufficiently large class of functions $k(x, x')$ one can prove the following two formulas:

$$k_\Gamma(x, x') = \sum_i h(\lambda^i) f_i(x) \overline{f_i(x')} \tag{11.5}$$

is the expansion in eigenfunctions (the bar denotes complex conjugation) and

$$\sum_i h(\lambda^i) = \int_F k_\Gamma(x, x)\, d\mu(x) = \sum_{\gamma \in \Gamma} \int_F k(x, \gamma x)\, d\mu(x) \tag{11.6}$$

is the trace formula.

By means of the transformation analogous to that used in Chapter 6 in the case of the upper half-plane, one can arrive from formula (11.6) at the Selberg trace formula:

$$\sum_i h(\lambda^i) = \sum_{\{\gamma\}_\Gamma} |\Gamma_\gamma \backslash G_\gamma| g(\{\gamma\}_G). \tag{11.7}$$

In formula (11.7) we denote by G_γ the centralizer of γ in G, by Γ_γ the centralizer of γ in Γ, by $|\Gamma_\gamma \backslash G_\gamma|$ the volume of the fundamental domain of Γ_γ in G_γ; by $\{\gamma\}_G$, $\{\gamma\}_\Gamma$ the conjugacy classes with the representative γ in G and Γ respectively; γ runs through the set of all conjugacy classes in Γ; the function g is defined making use of the function h by means of the Harish–Chandra transformation, in particular.

We now interpret the obtained Selberg trace formula from a representation theory standpoint. First we recall some definitions from Lie group theory. Let G be a Lie group, i.e., G be the group which is simultaneously an analytic manifold, and the group operations be analytic functions, as well. The group G will be called simple if it does not contain any closed connected normal divisor, and it will be called semi-simple if it does not contain any closed connected commutative normal divisor. Finally, G is reductive if its factor group by the center is semi-simple. One can obtain the symmetric space X_0 if we consider the factor space $X_0 \simeq G/K$, where G is semi-simple, K is a maximal compact subgroup of G (see Chapter 13).

Suppose that G is a reductive Lie group, and we choose as X the group G itself. As before, let Γ be a discrete subgroup of G with compact factor $\Gamma\backslash G$. Let $\hat{G}$ denote the set of unitary equivalent classes of irreducible unitary representations of the group G. The spectral expansion (11.5) turns out to be closely connected with the decomposition into irreducible representations $U \in \hat{G}$ of the regular representation of G in $L_2(\Gamma\backslash G; d\mu)$.

In fact, let $R_{\Gamma\backslash G}$ be a regular (quasi-regular) representation of the group G in the space $L_2(\Gamma\backslash G; d\mu)$ (i.e. the representation by the shift operators $f(x) \to f(xg)$, $g \in G$).

For any function $\alpha \in C_0^\infty(G)$ we have

$$R_{\Gamma\backslash G}(\alpha) = \int_G \alpha(x) R_{\Gamma\backslash G}(x)\, d\mu(x). \tag{11.8}$$

The linear operator (11.8) is the integral operator in $L_2(\Gamma\backslash G; d\mu)$ with the kernel

$$k_\alpha(x, x') = \sum_{\gamma\in\Gamma} \alpha(x^{-1}\gamma x'). \tag{11.9}$$

From the choice of the function α and from the compactness of $\Gamma\backslash G$ it follows that $R_{\Gamma\backslash G}(\alpha)$ is the operator of the Hilbert–Schmidt type and, thus, it is compact.

According to the classical theorem of representation theory of a Lie group in a separable Hilbert space, the condition of compactness of the operators $R_{\Gamma\backslash G}(\alpha)$ implies the existence of the desired decomposition of the representation $R_{\Gamma\backslash G}$ into the direct sum of unitary irreducible representations, each of which enters with finite multiplicity, i.e.:

$$R_{\Gamma\backslash G} = \sum_{U\in\hat{G}} \oplus m(U;\Gamma)U. \tag{11.10}$$

Now follows the method of deriving the Selberg trace formula from (11.10). It is clear a priori that among the functions α generating the operators $R_{\Gamma\backslash G}(\alpha)$ there exist the functions for each of which the operator $R_{\Gamma\backslash G}(\alpha)$ is of trace class. Furthermore, from the Malliavin–Dixmier theorem it follows that any function $\alpha \in C_0^\infty(G)$ can be written in the form of a finite sum of convolutions $\alpha_i * \beta_i$, where α_i, $\beta_i \in C_0^\infty(G)$. Hence, it follows that $R_{\Gamma\backslash G}(\alpha)$ can be written as a finite sum of the products of the operators of the type of Hilbert–Schmidt operators and, consequently, it is of trace class. So, the spectral trace of $R_{\Gamma\backslash G}(\alpha)$ coincides with the matrix trace

$$\sum_{U\in\hat{G}} m(U;\Gamma) T_U(\alpha) = \int_{\Gamma_U\backslash G} k_\alpha(x, x')\, d\mu(x), \tag{11.11}$$

where T_U is the character of the representation U.

One can transform the right-hand side of (11.11) in accordance with (11.9) to the form containing so-called orbital integrals. The Selberg trace formula takes the form (compare with (11.7))

$$\sum_{U\in\widehat{G}} m(U;\Gamma)T_U(\alpha) = \sum_{\{\gamma\}_\Gamma} |\Gamma_\gamma\backslash G_\gamma| \int_{G_\gamma\backslash G} \alpha(x^{-1}\gamma x)\, d\mu_{G_\gamma\backslash G}(x), \tag{11.12}$$

where $d\mu_{G_\gamma\backslash G}$ is the measure on $G_\gamma\backslash G$. In the formula (11.12) the function α enters the right-hand and left-hand part disparately. To restore the balance, the orbital integrals are subjected to Fourier transformation in the spirit of Harish–Chandra. We shall not cite here the results obtained in this direction, referring the reader instead to corresponding literature (see Chapter 13).

Now to return to the beginning of this chapter in order to consider a much more difficult and interesting aspect of the theory when the discrete group Γ has a non-compact fundamental domain and $\mu(F) < \infty$. Each of the operators $K \in V$ in $L_2(F;d\mu)$ has a continuous spectrum. Properties of the continuous spectrum are determined by a geometry of the fundamental domain F. As usual, the canonical fundamental domain is constructed starting from a certain base point $x_0 \in X$, and it contains those and only those points $x \in X$ for which $d(x,x_0) \le d(x,\gamma x_0)$ for all $\gamma \in \Gamma$, where d is the geodesic distance.* Since F is non-compact then F contains at least one geodesic ray starting at the point x_0 and having an infinite length. In contrast to the simple case of the upper half-plane considered earlier (see Chapter 4), in the general case the set of these rays can contain both the isolated rays and the rays filling out the surfaces of various dimensions. This circumstance defines the properties of the continuous spectrum of the problem.

We define as before the operator K_Γ in $L_2(F;d\mu)$ using the operator $K \in V$. From the general theorem of functional analysis it follows the existence of the spectral measure $d\nu_\Gamma(\lambda)$ such that

$$K_\Gamma = \int h(\lambda)\, d\nu_\Gamma(\lambda), \tag{11.13}$$

where the integration is over the spectrum of the ring V in $L_2(F;d\mu)$. Unlike the formula (11.5) written for the co-compact group Γ, the measure $d\nu_\Gamma(\lambda)$ in the considered case also has the continuous part $d\nu_\Gamma^c(\lambda)$. For its effective description a system of Eisenstein series is defined, each of which is then continued analytically in the neighbourhood of the spectrum, where these series determine the measure $d\nu_\Gamma^c$.

We now define, according to Selberg, the Eisenstein series connected with the continuous spectrum and outline a method of their analytic continuation in the case when X is a symmetric space. At first, we must consider the simplest variant of the theory when the given fundamental domain $F \cong \Gamma\backslash X$ contains only finitely many geodesic rays of infinite length passing through the base point. The discrete groups of such type act in the spaces of any rank. Recall that $X = G/K$, where G

* We permit here a small evident inaccuracy relative to boundary points of F.

is a semi-simple Lie group, $K \subset G$ is a maximal compact subgroup. The rank of X is the dimension of the split torus A in the Iwasawa decomposition $G = NAK$, N is a unipotent subgroup (the rank also coincides with the number of generators of the ring V). We now give a coordinate definition of the Eisenstein series (see [165]).

There exists a parabolic subgroup $P_1 \subset G$ of rank one, i.e., a subgroup which can be written as a product of the groups $P_1 = A_1 M_1 N_1$, where M_1 is semi-simple, N_1 is unipotent, A_1 is a one-dimensional (real) torus, and which acts transitively on X. We introduce by means of P_1 the coordinates on $X : x \in X$, $x = (a, m, n)$, where $a \in A_1$, $m \in M_1/K_1$, $n \in N_1$, K_1 is a maximal compact subgroup of M_1. The invariant Riemann metric in X and corresponding Riemann measure in these coordinates have the form

$$ds^2 = c_1 \frac{da^2}{a^2} + c_2\, d\tilde{s}^2 + Q_{a,m,n}(dn), \quad d\mu(x) = c_3 \frac{da}{a^2}\, dm\, dn$$

(we choose the coordinate $a \in A_1$ by a special way), where $c_1, c_2, c_3 = \text{const}$, $d\tilde{s}^2$ is the Riemann metric on M_1/K_1, Q is a positive definite quadratic form of dn, the coefficients of which vanish as $a \to \infty$.

We consider the intersections $\Gamma_P = \Gamma \cap P_1$, $\Gamma_N = \Gamma \cap N_1$, $\Gamma_1 = \Gamma \cap M_1$. Suppose that the factors N_1/Γ_N and M_1/Γ_1 are compact. Then the group Γ has the fundamental domain with necessary properties. So if $X_{a_0} = \{x \in X \mid a \geq a_0\}$ and $F_{a_0} = F \cap X_{a_0}$, then for a sufficiently large a_0, F_{a_0} is the fundamental domain of Γ_P in X_{a_0}. In addition, from any point $x \in F_{a_0}$ one can draw only one geodesic ray of infinite length lying entirely in F_{a_0}.

Thus, Γ_P defines as $a = \infty$ the point-like cusp or the vertex of type one of the fundamental domain F of the group Γ. Suppose that Γ has only one cusp of type one.

We call the following series Eisenstein series

$$E(x; s; v) = \sum_{\gamma \in \Gamma_P \backslash \Gamma} f_{s,v}(\gamma x), \tag{11.14}$$

where $f_{s,v}(x) = a^s v(m)$, and $v(m)$ is an invariant relative to Γ_1 eigenfunction of basis operators $\mathcal{D}_i'$ on the symmetric space M_1/K_1. The series (11.14) is absolutely convergent for $\operatorname{Re} s > 1$ and defines a Γ-automorphic eigenfunction of the operators $\mathcal{D}_i$ on X depending analytically on s.

The series (11.14) admits a meromorphic continuation into the region $\operatorname{Re} s \leq 1$. The proof is analogous to that considered in Chapter 4 and is based on the Maass–Selberg relation. The integral

$$\frac{1}{|\Gamma_N \backslash N_1|} \int_{\Gamma_N \backslash N_1} E(nx; s; v)\, dn \tag{11.15}$$

which is equal to $a^s v(m) + \varphi(s) w(m) a^{1-s}$, where $w(m)$ is the function of the same type as $v(m)$, plays the role of the constant term. Meromorphic continuation of

$E(x;s;v)$ follows from meromorphy of $\varphi(s)$, functional equation $\varphi(s)\overline{\varphi(1-\bar{s})}=1$ functional equation

$$E(x;s;v)=\varphi(s)E(x;1-s;w). \tag{11.16}$$

The situation is intentionally simplified. In fact, the proof contains many details (see [165], [123], [124]). In particular, formula (11.16) is valid if the eigenfunction $f_{s,v}(x)$ is of multiplicity 1. If there exist several functions v for which $f_{s,v}(x)$ belong to the same eigenvalue then instead of (11.16) there exists the equation showing that the system $E(x;s;v)$ is connected with the system $E(x;1-s;v)$ by means of the matrix $\Phi_\Gamma(s)$. This matrix becomes unitary for $\operatorname{Re} s=\frac{1}{2}$.

Formula (11.16) is generalized to the case where F has finitely many vertices of type one. Each function $E(x;\frac{1}{2}+ir;v)$ obtained for different vertices and different v is an eigenfunction of the continuous spectrum of the ring V in $L_2(F;\,d\mu)$.

The problem becomes much more complicated if we try to construct the full set of eigenfunctions of the continuous spectrum of the operators for a discrete group Γ a fundamental domain of which has arbitrary (not necessarily type one) cusps. In particular, Eisenstein series on many complex variables and Eisenstein series generated by parabolic forms on a space of lesser rank appear. We now consider a non-trivial example of constructing such series for $G=\mathrm{SL}(3,\mathbb{R})$, $\Gamma=\mathrm{SL}(3,\mathbb{Z})$ which leads to the full set of eigenfunctions of the continuous spectrum. Next, following Langlands, we consider the general case.

Introduce the following parabolic subgroups of the group G:

$$P_0=\left\{\begin{pmatrix} a_{11} & a_{12} & a_{13}\\ 0 & a_{22} & a_{23}\\ 0 & 0 & a_{33}\end{pmatrix}\right\},\quad P_1=\left\{\begin{pmatrix} a_{11} & a_{12} & a_{13}\\ a_{21} & a_{22} & a_{23}\\ 0 & 0 & a_{33}\end{pmatrix}\right\},$$

$$P_2=\left\{\begin{pmatrix} a_{11} & a_{12} & a_{13}\\ 0 & a_{22} & a_{23}\\ 0 & a_{32} & a_{33}\end{pmatrix}\right\}.$$

For each group P_j we construct the Langlands factorization $P_j=N_jA_jM_j$, where N_j is unipotent, A_j is splittable, M_j is simple for $j=0,1,2$. The fundamental domain F of the group Γ has only one cusp which has a more complicated structure than the vertex of type one. In particular, it contains continuum geodesic rays which go out from a base point and is described by two subgroups $\Gamma\cap P_2$ and $\Gamma\cap P_1$. We define the Eisenstein series thus:

$$\begin{aligned} E_0(x;\lambda)&=\sum_{\gamma\in\Gamma\cap P_0\backslash\Gamma}\exp(\lambda(H_0(\gamma x)))+\rho(H_0(\gamma x)),\\ E_j(x;v_k;\lambda)&=\sum_{\gamma\in\Gamma\cap P_j\backslash\Gamma}[\exp(\lambda(H_0(\gamma x)))+\rho(H_0(\gamma x))]v_k(\gamma x),\quad j=1,2,\end{aligned} \tag{11.17}$$

using the following notation: $\exp H_j(x)=a_j(x)=a_j$, in the factorization $x=n_ja_jm_j$, where $n_j\in N_j$, $a_j\in A_j$, $m_j\in M_j/M_j\cap K$, ρ is a half-sum of the positive roots, see Chapter 13, $\rho = \frac{1}{2}(\alpha_1 + \alpha_2 + \alpha_3)$, λ is a linear function on a complex span of the algebra $\log A_j$. In addition, $\{v_k\}$, $k = 1, 2, \ldots$ is the full set of eigenfunctions of the discrete spectrum of the automorphic Laplacian for $M_j/M_j \cap K$ in the corresponding Hilbert space.

Each of the Eisenstein series converges in its region λ given by the inequalities $(\operatorname{Re}\lambda, \alpha_l) > (\rho, \alpha_l)$, $l = 1, 2, 3$, and is analytic in λ there. To construct the spectral expansion, it is necessary to continue all of the series analytically to the spectrum $\operatorname{Re}\lambda = 0$. The structure of 'higher' Eisenstein series is simplest. Its meromorphy follows from meromorphy of the following set of functions giving the functional equations for it:

$$c(t;\lambda) = \prod_{\alpha_j > 0, t\alpha_j < 0} \frac{\xi(\lambda(H^0_{\alpha_j}))}{\xi(\lambda(H^0_{\alpha_j}) + 1)},$$

where $\xi(t) = \pi^{-1/2}\Gamma(t/2)\varsigma(t)$, $\varsigma(t)$ is the Riemann zeta-function, $t \in W$ is an element of the Weyl group for G (see Chapter 13), $\lambda(H^0_{\alpha_j}) = (\lambda, \alpha_j)$,

$$E_0(x;\lambda) = c(t;\lambda)E_0(x;t\lambda).$$

One can obtain similar assertions for other series (11.17) although their nature is more transcendental (especially for the series $E_j(x; v_k; \lambda)$ for parabolic forms v_k). Nevertheless, each of those series admits meromorphic continuation and has the functional equations:

$$E_1(x; c_2(v;\lambda); t_1\lambda) = E_2(x; v; \lambda), \quad E_2(x; c_1(v;\lambda); t_2\lambda) = E_1(x; v; \lambda),$$

where $t_1, t_2 \in W$, $c_1(\,\cdot\,;\lambda), c_2(\,\cdot\,;\lambda)$ are linear operators in the space of functions v_k depending meromorphically on λ and satisfying the functional equations

$$c_1(c_2(v;\lambda); t_2\lambda) = v, \quad c_2(c_1(v;\lambda); t_1\lambda) = v, \quad v = v_k.$$

Allthough all the results mentioned above can be obtained by special methods, they are all the consequences of general theory of the Eisenstein series developed by Langlands, which is at least valid for any reductive group G and its arithmetic discrete subgroup Γ. To make precise assertions on this theory, it is necessary to return to a viewpoint of representation theory of Lie groups.

So, we have the regular representation $R_{\Gamma\backslash G}$ where Γ is an arithmetic subgroup of G with a non-compact factor $\Gamma\backslash G$ of a finite volume. The problem consists, as before, in decomposition of this representation into irreducible ones. The operator $R_{\Gamma\backslash G}(\alpha)$ defined by formula (11.8) ceases to be not only the operator of Hilbert–Schmidt type, but also compact, which means the existence of a continuous spectrum in addition to a discrete one.

The Hilbert space $L_2(\Gamma\backslash G; d\mu)$ decomposes into a direct sum of invariant subspaces:

$$L_2(\Gamma\backslash G; d\mu) = L_2^{\text{dis}}(\Gamma\backslash G; d\mu) \oplus L_2^{\text{con}}(\Gamma\backslash G; d\mu). \tag{11.18}$$

To construct a generalized Selberg trace formula, we must first examine efficiently these two spaces of discrete and continuous spectrum. Then it is necessary to separate the discrete part of the representation, to decompose it into irreducible representations and to calculate the trace of a projection. But this program is very difficult and is not realized at present.

Now we return to the Eisenstein series which are needed for an efficient description of the subspace L_2^{con} and partially L_2^{dis}.

Suppose that Γ is the arithmetic subgroup of the group G of real points of the reductive group with one cusp in the fundamental domain. We shall define cusp form as an automorphic relative to Γ function on G for which the integral of type (11.15) vanishes for a unipotent radical N of any parabolic subgroup $P \subset G$; $P = AMN$, A is splittable, M is similar to G. On the set of parabolic subgroups of G, there exists an equivalence relation which we shall call the associativity relation. Two subgroups are associated if, loosely speaking, the complex envelopes of algebras of their split tori can be transferred one into another by an element of the Weyl group. Let $\{P\}$ be a class of the subgroups associated to P in G.

The space $L_2(\Gamma\backslash G; d\mu)$ decomposes into the following direct sum of invariant subspaces

$$L_2(\Gamma\backslash G; d\mu) = L_2^{\text{cusp}}(\Gamma\backslash G; d\mu) \oplus \sum_{\{P\}} \oplus L_2(\Gamma\backslash G; d\mu; \{P\}), \tag{11.19}$$

the summation is taken over non-trivial classes of associated subgroups P, L_2^{cusp} is the subspace of cusp forms. The subspace $L_2(\Gamma\backslash G; d\mu; \{P\})$ is defined by the class of associated parabolic subgroups. It can be described as follows. Each element $\widehat{\Phi}(g) \in L_2(\Gamma\backslash G; d\mu; \{P\})$ (more precisely, an element of a dense set) can be obtained by the Fourier transform

$$\widehat{\Phi}(g) = \int_{\operatorname{Re}\Lambda=\Lambda_0} \sum_{\gamma\in\Gamma\cap P\backslash\Gamma} [\exp(\Lambda(H(\gamma g)) + \rho(H(\gamma g)))]\Phi(\Lambda;\gamma g)\, d\Lambda \tag{11.20}$$

for the corresponding function $\Phi(\Lambda; g)$. From (11.20) it is natural to define the general Eisenstein series by the formula

$$E(g; \Phi; \Lambda) = \sum_{\Gamma\cap P\backslash\Gamma} \Phi(\gamma g) \exp(\Lambda(H(\gamma g)) + \rho(H(\gamma g))) \tag{11.21}$$

(the notation is the same as in the example with the group $\mathrm{SL}(3, \mathbb{R})$). The function Φ is taken in a special class. The series (11.21) is defined for the Λ such that $(\operatorname{Re}\Lambda, \alpha) > (\rho, \alpha)$ relative to all positive roots α.

Formula (11.20) is at the basis of the following important formula for the inner product of the functions $\widehat{\Phi} \in L_2(\Gamma\backslash G; d\mu; \{P\})$, $\widehat{\Psi} \in L_2(\Gamma\backslash G; d\mu; \{P'\})$

$$\int_{\Gamma\backslash G} \widehat{\Phi}(g)\overline{\widehat{\Psi}}(g)\, d\mu(g) = \int_{\operatorname{Re}\Lambda=\Lambda_0} \sum_{t\in\Omega(\log A, \log A')} (c(t;\Lambda)\Phi(\Lambda), \Psi(-t\overline{\Lambda}))\, d\Lambda, \tag{11.22}$$

where Ω is the set of linear maps defined by elements of the Weyl group between the complex envelopes of the algebras $\log A$, $\log A'$ of the subgroups $P = AMN$, $P' = A'M'N'$, $c(t;\lambda)$ is the linear operator in the space of admissible functions Φ.

Before discussing formula (11.22), we note that the mentioned regular representation has only a discrete spectrum in the space L_2^{cusp}. This assertion is the known theorem of Gelfand and Pjatetskiĭ–Shapiro (see [95]). So, to solve the global problem of decomposition of $R_{\Gamma\backslash G}$ into irreducible representations, it suffices to consider a projection of $R_{\Gamma\backslash G}$ on each subspace $L_2(\Gamma\backslash G; d\mu; \{P\})$. Formula (11.22) is a necessary tool for such investigation. In fact, it may be interpreted as a formula of the type of Parseval formula for the desired spectral decomposition. In order to compute the spectral measure explicitly it is necessary to continue meromorphically the integrand in the right-hand side of (11.22) to at least $\operatorname{Re}\Lambda = 0$ for each pair of admissible functions $\Phi(\Lambda)$, $\Psi(\Lambda)$ (the right-hand side must have the form of a sum of the integral over the spectrum $\operatorname{Re}\Lambda = 0$ and a certain series of residues).

So, we arrive at the problem of analytic continuation of each operator $c(t;\Lambda)$ and Eisenstein series $E(g;\Phi;\lambda)$ for a certain admissible class of functions Φ from the region $(\operatorname{Re}\Lambda, \alpha) > (\rho, \alpha)$ onto the entire space Ξ conjugate to the complex envelope of $\log A$. The problem is solved by Langlands in his Main Theorem. A not very rigorous statement of the theorem follows (for a precise statement and the proof, see [124], [123]).

THEOREM 11.1 (1) *For any values of parameters each operator* $c(\,\cdot\,;\Lambda)$ *and the function* $E(\,\cdot\,;\,\cdot\,;\Lambda)$ *are meromorphic in the corresponding space* Ξ *and are analytic on the spectrum* $\operatorname{Re}\Lambda = 0$.

(2) *The following functional equations are valid:*

$$c(t_1t_2;\Lambda) = c(t_1;t_2\Lambda)c(t_2;\Lambda), \quad t_1,t_2 \in W,$$
$$E(g;c(t;\Lambda)\Phi;t\Lambda) = E(g;\Phi;\Lambda).$$

The proof of the theorem is long and technically difficult. By induction on the rank of a parabolic subgroup, one can reduce the statement of the theorem to meromorphic continuation of the Eisenstein series of one complex variable which corresponds to the rank one case. Langlands proved the last assertion, developing the ideas of Selberg mentioned in part above.

A discussion of the results concerning the derivation of the general Selberg trace formula for the Γ with non-compact factor of finite volume now follows.

Unfortunately, at present there is no exact classical analog of co-compact version of the Selberg trace formula in considered general situation. In fact, there is no such formula even for simplest discrete subgroups of rank greater than one (i.e., with vertices of the fundamental domain F of not type one), for $\Gamma = \mathrm{SL}(3,\mathbb{Z})$, for example. Although in the latter case the part of the spectral measure corresponding to the continuous spectrum was computed exactly (see [4], [3], the first published papers on this subject).

The most general results on the Selberg trace formula for a group of arbitrary rank are due to Arthur. In his papers (see [40–44]) he studied the regular representation of the adel group $R_{G(\mathbb{Q})\backslash G(\mathbb{A})}$, where G is a reductive matrix group defined over $\mathbb{Q}$. (A more detailed discussion on the adel group and its representations in connection with applications to L-functions appears in the following chapter.) The problem of decomposing this representation into irreducible ones contains the problem of decomposing the representation $R_{\Gamma(N)\backslash G}$ into irreducible ones for any congruence subgroup $\Gamma(N)$ of the group of integer points of G. Developing the ideas of Selberg, Langlands, Gelfand, Pjatetskiĭ–Shapiro, Borel, Jacquet, Duflo, and Labesse, Arthur transferred the theory of the Eisenstein series on adel groups and, in particular, proved the Main Theorem generalizing Theorem 11.1. Furthermore, he succeeded in constructing an analog of the Selberg trace formula in this general case. This is a brief outline of his results.

Introduce the notation $R = R_{G(\mathbb{Q})\backslash G(\mathbb{A})}$. Instead of the regular representation R, Arthur considers a 'twisted' representation of the group $G(\mathbb{A})$ in $L_2(AG(\mathbb{Q})\backslash G(\mathbb{A}))$. Then, the operator $R^{\xi}(\alpha)$, which is defined by means of the formula anlogous to (11.8) by the function α^{ξ} with a compact support, is an integral operator with the kernel

$$k^{\xi}_{\alpha}(x,x') = \sum_{\gamma\in G(\mathbb{Q})} \alpha^{\xi}(x^{-1}\gamma x').$$

The projection into the cusp-form space $R^{\xi}_{\text{cusp}}(\alpha)$ is given by the formula

$$k^{\xi}_{\alpha\,\text{cusp}}(x,x') = k^{\xi}_{\alpha}(x,x') - k^{\xi}_{\alpha E}(x,x'), \tag{11.23}$$

where $k^{\xi}_{\alpha E}$ is a sum over all non-trivial classes of associated parabolic subgroups P and over admissible functions Φ of integrals of the Eisenstein series taken along the spectrum (see [41]). Then, calculating the trace of $R^{\xi}_{\text{cusp}}(\alpha)$ in accordance with (11.23), Arthur arrives at the formula

$$\operatorname{Tr} R^{\xi}_{\text{cusp}}(\alpha) = \sum_{\chi} J_{\chi}(\alpha^{\xi}). \tag{11.24}$$

In the right-hand part of (11.24) there is the sum of special distributions defined firstly in a certain non-invariant way according to a parameter of the section of the factor $G(\mathbb{Q})\backslash G(\mathbb{A})$, which is connected with regularization of the trace (11.23).

Arthur called formula (11.24) a 'generalized Selberg trace formula'. As an interesting example, he considers the group GL(3) for which he succeeded in calculating the right-hand part of (11.24) for special functions α in terms of orbital integrals (see [41])

$$\operatorname{Tr} R^{\xi}_{\text{cusp}}(\alpha) = \sum_{\{\gamma\}_{G(\mathbb{Q})}} |AG_{\gamma}(\mathbb{Q})\backslash G_{\gamma}(\mathbb{A})| \int_{G_{\gamma}(\mathbb{A})\backslash G(\mathbb{A})} \alpha^{\xi}(x^{-1}\gamma x)\, d\mu(x).$$

CHAPTER 12

Automorphic Functions, Representations and L-Functions

In this chapter, we define certain L-functions playing an important role in modern number theory and try to clarify a connection between them and automorphic functions. At present, the general viewpoint on the wide range of problems connected with L-functions is due to the work of many mathematicians. This point of view can be observed especially clearly in the brilliant survey by Langlands, one of the founders of the theory as a whole [125], whose ideas we shall discuss in this chapter.

There are two essentially different ways of defining L-functions. The first gives L-function as a Euler product (in the region of its absolute convergence) the coefficients of which are defined by arithmetic formulas. In the second case L-function is defined, as a rule, as a meromorphic function on $\mathbb{C}$ satisfying the special functional equation. The (optimistic) general point of view assumes that the number of ways of determining L-functions is greater than L-functions themselves. Coincidences which occur are a realization of a general law which one should call the general reciprocity law. In such a way, one must answer traditional questions of L-function theory such as: the question of analytic continuation of given Dirichlet series, the question of existence of a functional equation for it and so on. In addition, the relations of the type of special reciprocity laws appear (see Chapter 10).

But there exist at present more conjectures than theorems in general L-function theory. We present here a more elementary part of obtained results including certain classical definitions not having a claim on completeness*. The reader can find a detailed description of the subject in [55].

L-functions first appear in the papers of Riemann and Dirichlet devoted to the problem of distribution of primes. If $q \in \mathbb{Z}_+$ and χ is the Dirichlet character modulo q then the series

$$L(s, \chi) = \sum_{n=1}^{\infty} \frac{\chi(n)}{n^s} \tag{12.1}$$

converges absolutely in the region $\operatorname{Re} s > 1$, and its sum is regular in this region.

---

* For a more professional survey on the Langlands program, see S. Gelbart, Bull. Amer. Math. Soc., 1 (1984), 2, 177–219.

The function (12.1) allows a meromorphic continuation on $\mathbb{C}$, and it is called the Dirichlet L-function. One of the most important properties of the Dirichlet L-function is that it can be given as a product over primes p,

$$L(s,\chi) = \prod_p \left(1 - \chi(p)p^{-s}\right)^{-1}, \quad \operatorname{Re} s > 1, \tag{12.2}$$

which is called a Euler product. It has another important property when it corresponds to a primitive character χ modulo q. Namely, it satisfies the functional equation

$$\xi(s,\chi) = \frac{\tau(\chi)}{i^a q^{1/2}} \xi(1-s,\overline{\chi}), \tag{12.3}$$

where

$$\xi(s,\chi) = \left(\frac{\pi}{q}\right)^{-1/2(s+a)} \Gamma\left(\frac{s+a}{2}\right) L(s,\chi),$$

$a = 0$, if $\chi(-1) = 1$ and $a = 1$, if $x(-1) = -1$;

$$\tau(\chi) = \sum_{n (\operatorname{mod} q)} \chi(n) e\left(\tfrac{n}{q}\right)$$

is the Gauss sum. In the case, where $q = 1$, $\chi = 1$, the function (12.3) is a Riemann ς-function.

From the point of view of algebraic number theory, the Dirichlet L-function is a special case of the Hecke L-function, and the Riemann ς-function is a special case of the Dedekind ς-function. Let $\mathbb{F}$ be a field of algebraic numbers, i.e. an extension of the field $\mathbb{Q}$ of finite degree. We now refer to some classical notions from algebraic number theory. For any point $v \in \mathbb{F}$, i.e. for any class of equivalent valuations, we denote the corresponding completion by $\mathbb{F}_v$. Let $\mathbb{A}$ or $\mathbb{A}_{\mathbb{F}}$ be a ring of adèles of the field $\mathbb{F}$. The elements of $\mathbb{A}$ are sequences $\{c_v\}$, enumerated by the points of the field $\mathbb{F}$ and satisfying the condition: for almost all of v (i.e. for all v except for a finite number of them) belonging to $c_v \in \mathcal{O}_v$ takes place where $\mathcal{O}_v$ is the ring of integers of the field $\mathbb{F}_v$, and for all v, $c_v \in \mathbb{F}_v$. $\mathcal{O}_v$ is the local ring defining for all non-Archimedian v. Let ω_v be a generator of its maximal ideal. If p is a prime ideal of the ring of integers of the field $\mathbb{F}$ and v is defined by p, i.e. v contains a p-valuation, then one often writes ω_p instead of ω_v. This element is also called a uniformizing parameter in p. The operations in $\mathbb{A}_{\mathbb{F}}$ are defined coordinate-wise by the operations in the fields $\mathbb{F}_v$. Lastly, we introduce the group $I_{\mathbb{F}}$, i.e. the group of idèles of the field $\mathbb{F}$, which is the group of invertible elements of the ring $\mathbb{A}_{\mathbb{F}}$.

The field $\mathbb{F}$ is included in $\mathbb{A}_{\mathbb{F}}$ and, in addition, $\mathbb{F}^*$ (the multiplicative group of the field $\mathbb{F}$) is included in $I_{\mathbb{F}}$: if $\alpha \in \mathbb{F}$, then $\alpha \to \{c_v\} \in A_{\mathbb{F}}$ with $c_v = \alpha$ for all v. The correctness of the definition follows from the fact that any element in $\mathbb{F}$ is an integer in almost all of $\mathbb{F}_v$. The field $\mathbb{F}_{v'}$ is included in $A_{\mathbb{F}}$ and, in addition, $\mathbb{F}^*_{v'}$

is included in $I_{\mathbb{F}}$ in the following way:

$$\text{if} \quad d \in \mathbb{F}_{v'} \quad \text{then} \quad d \to \{c_v\}$$

with $c_{v'} = d$, $c_v = 1$ for $v \neq v'$.

The images of $\mathbb{F}$ and $\mathbb{F}_v$ in $\mathbb{A}_{\mathbb{F}}$ are the subfields in $\mathbb{A}_{\mathbb{F}}$; the images of $\mathbb{F}^*$ and $\mathbb{F}_v^*$ in $I_{\mathbb{F}}$ are subgroups in $I_{\mathbb{F}}$. So, let $I_{\mathbb{F}}$ be the idèle group of the field $\mathbb{F}$ and χ be a character of the group $\mathbb{F}^*\backslash I_{\mathbb{F}}$. If v is the point of the field $\mathbb{F}$ and $\mathbb{F}_v$ its corresponding completion, then $\mathbb{F}_v^*$ is embedded in $I_{\mathbb{F}}$ and defines the character χ_v of the group $\mathbb{F}_v^*$. More strictly χ_v is a composition

$$\mathbb{F}_v^* \to I_{\mathbb{F}} \to \mathbb{F}^*\backslash I_{\mathbb{F}} \xrightarrow{\chi} \mathbb{C}^*$$

of the inclusion $\mathbb{F}_v^*$ in $I_{\mathbb{F}}$, the canonical projection $I_{\mathbb{F}} \to \mathbb{F}^*\backslash I_{\mathbb{F}}$ and of the character χ. The Hecke L-function is defined as a product

$$L(s,\chi) = \prod_v L(s,\chi_v) \tag{12.4}$$

over all points v. The factors $L(s,\chi_v)$ for Archimedian v consist of Γ-function. If v is a non-Archimedian point defined by a prime ideal p, and the character χ is trivial on units and this takes place for almost all v, then

$$L(s,\chi_v) = \left\{1 - \chi_v(\omega_p)(N_p)^{-s}\right\}^{-1},$$

$\omega(p)$ is the uniformizing parameter in p, N is the norm. Hecke L-functions are defined (12.4) in a certain half-plane of $\mathbb{C}$; they have a meromorphic continuation onto all of $\mathbb{C}$ and satisfy the functional equation of the form

$$L(s,\chi) = \epsilon(s,\chi)L(1-s,\chi^{-1}), \tag{12.5}$$

where $\epsilon(s,\chi)$ is the factor that can be calculated explicitly, see [95].

In the special case when $\mathbb{F} = \mathbb{Q}$, the Hecke L-functions coincide with the functions ξ from (12.3). In other words, the Dirichlet L-functions are the Hecke L-functions up to a factor corresponding to an Archimedian point v.

The more general construction belongs to Artin. Let $\mathbb{F}'$ be a Galois extension of the field $\mathbb{F}$ and

$$\sigma : \mathrm{Gal}(\mathbb{F}'/\mathbb{F}) \to \mathrm{GL}(n,\mathbb{C}) \tag{12.6}$$

is n-dimensional complex representation of the Galois group. For any point v of the field $\mathbb{F}$ we denote by σ_v the restriction of σ on the factorization group of the Galois group $\mathrm{Gal}(\mathbb{F}'/\mathbb{F})$ in v. An Artin L-function corresponding to σ is defined by the Euler product

$$L(s,\sigma) = \prod_v L(s,\sigma_v) \tag{12.7}$$

over all points v. If v is defined by the prime ideal p and the p is unramified in $\mathbb{F}'$ then

$$L(s,\sigma_v) = \frac{1}{\det(I - \sigma(\mathrm{Fr}_p)N(p)^{-s})}, \tag{12.8}$$

where Fr_p is the Frobenius element over p, I is the $n \times n$ unit matrix. The Artin L-function is defined by equalities (12.7), (12.8) for $s \in \mathbb{C}$ with a sufficiently large $\mathrm{Re}\, s$. The Artin conjecture is the following: for irreducible non-trivial representations σ the Artin L-functions can be continued to integer functions on $\mathbb{C}$. Artin proved his assumption for monomial representations σ, i.e. for such representations which are induced by one-dimensional representations of subgroups. The proof uses essentially the coincidence of Artin L-functions for one-dimensional σ with Hecke L-functions which were defined and studied earlier. In a general case, Brauer proved that Artin L-functions can be continued meromorphically onto $\mathbb{C}$.

Algebraic geometry supplies the more general construction bringing in correspondence certain L-functions to non-singular projective manifolds over number fields. If V is a non-singular projective manifold over an algebraic number field $\mathbb{F}$, then for almost all points v, the V has a good reduction over the residue field $\mathbb{F}_p$ where p is the ideal corresponding to v. For such p one can define a number $r(n)$ of points on reduced manifold with coordinates in the extension of the degree n over the field $\mathbb{F}_p$. Following Weil [183], we define the function

$$Z_p(s,V) = \exp\Big\{\sum_{n=1}^{\infty} \frac{r(n)}{n(Np)^{ns}}\Big\}.$$

It is acknowledged, mainly due to the papers by Dwork, Grothendieck, and Deligne, that

$$Z_p(s,V) = \prod_{i=0}^{2d} L_p^i(s,V)^{(-1)^i},$$

where d is the dimension of V, and $L_p^i(s,V)^{-1}$ is a polynomial of degree b_i of $(Np)^{-s}$, b_i is the ith Betti's number, so

$$L_p^i(s,V) = \prod_{j=1}^{b_i} \Big(1 - \frac{\alpha_{ij}(p)}{(Np)^s}\Big)^{-1}.$$

In addition, $|\alpha_{ij}(p)| = (Np)^{i/2}$.

Now one can consider the product

$$L^i(s,V) = \prod_v L_v^i(s,V) \tag{12.9}$$

over all points v ($L_v^i = L_p^i$ for the ideal p corresponding to v). It is assumed that the functions $L^i(s,V)$ can be continued meromorphically, and they satisfy the functional equations (similar to (12.6) given above) under the condition that the factors $L^i_v(s, V)$ in (12.9) are defined in a proper manner for infinite points v and for points v not having a good reduction.

In certain cases, the functions (12.9) can be represented as a product of Hecke L-functions. But these are particular cases. It is generally very difficult to study the functions (12.9) and this leads one to investigate the so-called standard L-functions which are defined in the theory of automorphic representations and generalize the Hecke L-functions.

There now follows a description of automorphic representations and standard L-functions.

First, we present a simple example. Consider a parabolic form f, $f \in \mathcal{H}_0(\Gamma; \chi)$; $\Gamma = \mathrm{SL}_2(\mathbb{Z})$, $\chi = 1$. Let $Af = \lambda f$ and

$$f(z) = \sum_{\substack{n \in \mathbb{Z} \\ n \neq 0}} \rho(n) \sqrt{y} K_{i\kappa}(2\pi |n| y) e(nx) \tag{12.10}$$

be a Fourier expansion ($x = \mathrm{Re}\, z$, $y = \mathrm{Im}\, z$, $z \in H$, $\kappa^2 = \lambda - \frac{1}{4}$ as in Chapter 5). It is easy to demonstrate (integrating (12.10) by terms) that

$$\int_0^\infty f(iy) u^{s-3/2}\, dy = 2^{-1} \pi^{-s} \Gamma\left(\frac{s+i\kappa}{2}\right) \Gamma\left(\frac{s-i\kappa}{2}\right) L_f(s), \tag{12.11}$$

where

$$L_f(s) = \sum_{\substack{n \in \mathbb{Z} \\ n \neq 0}} \frac{\rho(n)}{|n|^s}. \tag{12.12}$$

The Dirichlet series (12.12) converges absolutely if $\mathrm{Re}\, s$ is sufficiently large. The estimate of Theorem 5.5, for example, ensures an absolute convergence for $\mathrm{Re}\, s > \frac{3}{2}$. From (12.11) we obtain that the function (12.12) can be continued holomorphically onto the whole complex plane $\mathbb{C}$ since $f(iy)$ decreases sufficiently rapidly as $y \to \infty$ and as $y \to 0$. Then, from invariance relative to the transformation

$$\begin{pmatrix} 0 & -1 \\ 1 & 0 \end{pmatrix} \in \mathrm{SL}_2(\mathbb{Z})$$

we have

$$f(iy) = f(iy^{-1})$$

for all $y \in \mathbb{R}_+$. Hence, it is easy to see that the integral in (12.11) is invariant relative to the change of s on $1 - s$ and, therefore, L_f satisfies the functional equation

$$\pi^{-s} \Gamma\left(\tfrac{s+i\kappa}{2}\right) \Gamma\left(\tfrac{s-i\kappa}{2}\right) L_f(s) = \pi^{-(1-s)} \Gamma\left(\tfrac{1-s+i\kappa}{2}\right) \Gamma\left(\tfrac{1-s-i\kappa}{2}\right) L_f(1-s). \tag{12.13}$$

Suppose in addition that f is an eigenfunction of all Hecke operators T_p and

$$\rho(n) = \rho(-n) \tag{12.14}$$

for all n. Then the relations described in Chapter 5 take place, and we have the factorization

$$L_f(s) = 2\rho(1)\prod_p L_{f,p}(s), \tag{12.15}$$

where the product is over all primes p and

$$L_{f,p}(s) = \rho(1)^{-1}\sum_{l=0}^{\infty}\frac{\rho(p^l)}{p^{ls}} = \frac{1}{\left(1-\frac{\alpha_p}{p^s}\right)\left(1-\frac{\alpha_p^{-1}}{p^s}\right)} \tag{12.16}$$

for certain $\alpha_p \in \mathbb{C}$, and $\alpha_p + \alpha_p^{-1} = \rho(p)\rho(1)^{-1}$.

So, by means of the Mellin transform (12.11) we bring to a parabolic form f the function L_f given by the Dirichlet series (12.12), having the functional equation (12.13) and the Euler product (12.15) like L-functions previously considered. Hecke [97] was the first to study systematically, by means of the Mellin transform, the connection between automorphic functions and the Dirichlet series. This idea can also be observed in Riemann's work (see the derivation of a functional equation for the ς-function in [175]). Hecke considered the regular parabolic forms. This example is taken from Maass [130].

One can realize a transfer to automorphic representations in the following way. Let f be as above or, more generally, let

$$f \in \mathcal{H}(\Gamma(N), 1),$$

$\Gamma(N)$ be the principal congruence subgroup mod N. For $\sigma \in \mathrm{GL}_2^+(\mathbb{Q})$ ($\mathrm{GL}_2^+(\mathbb{Q})$ is a subgroup in $\mathrm{GL}_2(\mathbb{Q})$ consisting of matrices with the positive determinant) suppose

$$f_\sigma(z) = f(\sigma z), \quad z \in H.$$

Let $\Omega(f)$ be a $\mathbb{C}$-linear space generated by all functions f_σ. A representation of the group $\mathrm{GL}_2^+(\mathbb{Q})$ in the space $\Omega(f)$ is defined naturally. This representation turns out to be an algebraic irreducible iff the Dirichlet series L_f factorizes into the Euler product or, similarly, f is the eigenfunction of all Hecke operators (see [15]). We define a topology on the group $\mathrm{GL}_2^+(\mathbb{Q})$ in order to obtain the topological group. To do this we take the set of all congruence subgroups $\Gamma(N)$ as a basis of a system of neighbourhoods of the unit. The group

$$\overline{\mathrm{GL}_2^+(\mathbb{Q})} = \left\{\sigma = \prod_p \sigma_p \in \prod_p \mathrm{GL}_2(\mathbb{Q}_p) \mid \det\sigma_p \in \mathbb{Q}_+\right\} \tag{12.17}$$

is the completion of $\mathrm{GL}_2^+(\mathbb{Q})$ in this topology ($\mathbb{Q}_p$ is the field of p-adic numbers; the product in (12.17) is taken over all primes p). The representation of $\mathrm{GL}_2^+(\mathbb{Q})$ in $\Omega(f)$ is such that each element of the space $\Omega(f)$ is invariant relative to a certain congruence subgroup (f is invariant relative to $\Gamma(N)$, and f_σ is invariant relative to $\sigma^{-1}\Gamma(N)\sigma \cap \Gamma(N) \supset \Gamma(M)$ for a certain M). This enables one to continue the considered representation to a representation of the group $\overline{GL_2^+(\mathbb{Q})}$ in $\Omega(f)$. We let π_f denote this.

Then, the representation π_f is equal to $\otimes_p \pi_{f,p}$, where $\pi_{f,p}$ is a representation of the group $GL_2(\mathbb{Q})$ and $\otimes_p$ denotes the restricted tensor product over all primes p (see [80]). Now it seems natural to correspond the representation $\pi_{f,p}$ with a function $L_{f,p}$ from (12.16), and to representation π_f the product of these functions, i.e. L_f from (10.15).

Jacquet and Langlands [109], [53] found that in this construction one can eliminate the given automorphic function f. Namely, they propose a certain way of corresponding to each irreducible admissible representation of the group $GL_2(\mathbb{Q}_p)$ a matrix

$$h_p = \begin{pmatrix} \alpha_p & 0 \\ 0 & \beta_p \end{pmatrix} \in GL_2(\mathbb{C}) \tag{12.18}$$

and, at the same time, a function

$$L_p(s, \pi_p) = \frac{1}{\det(I - h_p p^{-s})} = \frac{1}{(1 - \alpha_p p^{-s})(1 - \beta_p p^{-s})}. \tag{12.19}$$

Let $\mathbb{A}$ be an adèle ring of the field $\mathbb{Q}$, let π be an irreducible admissible representation of the adel group $GL_2(\mathbb{A})$ and let

$$\pi = \bigotimes_p \pi_p \tag{12.20}$$

be a decomposition into the restricted tensor product over all the points, i.e. over all primes p including $p = \infty$. We correspond with a representation π a function

$$L(s, \pi) = \prod_p L_p(s, \pi_p) \tag{12.21}$$

i.e., the product of the functions (12.19). The construction of Jacquet and Langlands is such that in a special case where there exists an automorphic function f for which $\pi_p = \pi_{f,p}$ in (12.20), we obtain the function L_f by formula (12.21) from (12.15) up to a factor corresponding to $p = \infty$. We denote by $GL_2(\mathbb{A})$ (not being too literal) the product of the groups $GL_2(\mathbb{Q}_p)$ over all p (including $p = \infty$) restricted by the subgroups $GL_2(\mathbb{Z}_p)$, $\mathbb{Z}_p$ is the ring of integers of the field $\mathbb{Q}_p$. So,

$$GL_2(\mathbb{A}) = \Big\{ \sigma = \sigma_\infty \prod_p \sigma_p \mid \sigma_\infty \in GL_2(\mathbb{R}); \sigma_p \in GL_2(\mathbb{Q}_p);$$
$$\sigma_p \in GL_2(\mathbb{Z}_p) \quad \text{for almost all} \quad p \Big\}.$$

Considering the group $GL_2(\mathbb{A})$ (and its representations) turns out to be preferable in some ways to considering the group $GL_2^+(\mathbb{Q})$.

The representation π from (12.20) is, in essence, an irreducible component of the natural representation of the group $GL_2(\mathbb{A})$ in the space of continuous functions on $GL_2(\mathbb{Q})\backslash GL_2(\mathbb{A})$. The space of functions on the homogeneous space

$\mathrm{GL}_2(\mathbb{Q})\backslash\mathrm{GL}_2(\mathbb{A})$ 'contains' all spaces $\mathcal{H}(\Gamma(N),1)$. More precisely, if f is a function on $\mathrm{GL}_2(\mathbb{A})$ which is invariant on the left relative to $\mathrm{GL}_2(\mathbb{Q})$ and on the right relative to a certain open compact subgroup in $\mathrm{GL}_2(\mathbb{A})$, then the restriction of f on $\mathrm{GL}_2(\mathbb{R})$ is a function invariant on the left relative to a certain congruence subgroup of $\Gamma(N)$.

So, we proceed from studying functions on the upper half-plane automorphic relative to congruence subgroups to studying functions on $\mathrm{GL}_2(\mathbb{Q})\backslash\mathrm{GL}_2(\mathbb{A})$ and representations of the group $\mathrm{GL}_2(\mathbb{A})$.

There is plenty of scope here for generalizations. It is clear that one can consider GL_n instead of GL_2 for an arbitrary n or, more generally, a connected reductive group G. One can take any global field $\mathbb{F}$ and its adel ring $\mathbb{A}_\mathbb{F}$ instead of the field $\mathbb{Q}$ and its adel ring $\mathbb{A}$.

An automorphic representation π of the group $\mathrm{GL}_n(\mathbb{A}_\mathbb{F})$ is an irreducible component of the natural representation of $\mathrm{GL}_n(\mathbb{A}_\mathbb{F})$ in the space of continuous functions on $\mathrm{GL}_n(\mathbb{F})\backslash\mathrm{GL}_n(\mathbb{A}_\mathbb{F})$. Each such representation decomposes into a restricted tensor product $\pi = \otimes_v \pi_v$ over all points v of the field $\mathbb{F}$. One brings in correspondence to the representation π an L-function

$$L(s,\pi) = \prod_v L(s,\pi_v), \tag{12.22}$$

where for a finite point v corresponding to a prime ideal p

$$L(s,\pi_v) = \prod_{i=1}^{n} \frac{1}{1-\alpha_i(p)/N(p)^s}, \tag{12.23}$$

and the matrix

$$A(\pi_v) = \begin{pmatrix} \alpha_1(p) & & 0 \\ & \ddots & \\ 0 & & \alpha_n(p) \end{pmatrix}$$

is invertible for almost all p. The functions in (12.22), (12.23) are called standard L-functions. They can be continued analytically onto $\mathbb{C}$ and satisfy the functional equation

$$L(s,\pi) = \epsilon(s,\pi)L(1-s,\tilde{\pi}), \tag{12.24}$$

where $\tilde{\pi}$ is the representation contragradient to π (see [54]).

If $n=1$, then $\mathrm{GL}_n(\mathbb{A}_\mathbb{F}) = I_\mathbb{F}$, $\mathrm{GL}_n(\mathbb{F}) = \mathbb{F}^*$, so the space $\mathrm{GL}_n(\mathbb{F})\backslash\mathrm{GL}_n(\mathbb{A}_\mathbb{F})$ coincides with $\mathbb{F}^*\backslash I_\mathbb{F}$. It is well-recognized that the standard L-functions coincide in the case $n=1$ with the Hecke L-functions (see Tate [173]). One can construct a more general class of automorphic L-functions if one models a transition from Hecke L-functions to Artin L-functions. Namely, if σ and $\mathbb{F}'$ are as in (12.6) then we may set

$$L(s,\pi,\sigma) = \prod_v L(s,\pi_v,\sigma_v), \tag{12.25}$$

where

$$L(s, \pi_v, \sigma_v) = \frac{1}{\det(I - \sigma(\mathrm{Fr}_p)N(p)^{-s})} \tag{12.26}$$

for a finite point v defined by the ideal p. These functions are investigated in some special cases, for example, when $\sigma = \mathrm{Sym}^k(\mathrm{St})$ is a kth symmetrical power of the standard representation $\mathrm{St} : \mathrm{GL}_2(\mathbb{C}) \to \mathrm{GL}_2(\mathbb{C})$, $k = 2, 3, 4$ (see [85], [55]).

According to Langlands' conjecture, Artin L-functions (12.8) coincide with L-functions (12.22) of irreducible parabolic representations π of the group $\mathrm{GL}_n(\mathbb{F})$. The proof of Langlands' conjecture together with Jacquet's investigations [92] would lend proof to Artin's conjecture (see Chapter 13).

CHAPTER 13

Remarks and Comments. Annotations to the Cited Literature

Chapter 2.

The classical theory of modular functions together with the applications to arithmetics of quadratic forms is described brilliantly in Serre [170]. The reader can also find there the proof of Theorem 2.1.

The current simplified proof of Theorem 2.2, together with the bibliography of classical papers on this subject is given in [168]. We would also like to recommend Shimura [171] which is very well-written and contains a great deal of information on the connection between the theory of automorphic functions, algebraic geometry and number theory. It is readily comprehensible for readers from differing disciplines.

Hadamard's survey [1] is a good introduction to Poincaré's classical paper. The current perspective and the connection with the uniformization problem are both in the survey [98].

Chapter 3.

Comparing the theorem of Hardy and Landau on the estimate $R(X) = \Omega_{\pm}(X^{1/4})$ and the result of Landau and Valfish on the order of a mean value for the integral

$$\int_0^X R^2(t)\,dt \underset{X\to\infty}{\sim} \text{const}\, X^{3/2}$$

suggests that one could estimate $R(X) = O(X^{1/4+\epsilon})$, (i.e., $\theta = 1/4$). On the other hand, in spite of major efforts, the best of the known estimates have the worst order $\theta = \frac{1}{3} - \delta$, $\delta > 0$. We indicate here only the intermediate results: $\delta \approx \frac{1}{300}$ (van der Corput), $\delta \approx \frac{1}{120}$ (Hua Loo-Keng), $\delta \approx \frac{1}{111}$ (Chen) see [62]).

The formulas similar to the Voronoĭ–Hardy formula are also used when one examines the divisor problem of Dirichlet; see, for example, [61].

Chapter 4.

We mention here three monographs [171], [131], [129] in which the reader can find both initial and in-depth information on discrete groups and automorphic functions on the Lobachevsky plane. It is also worth mentioning the useful survey [174] written in a more 'physical' style.

For an alternative definition of the Fuchsian group of the first kind, see [129].

As for Selberg's proof of the spectral theorem, one must first examine his papers [163–165] which are now considered classics. The papers by Gelfand, Langlands, Harisch–Chandra, Godement, Roelcke, Elstrodt, Kubota, Neunhoffer, etc. played a significant role in the understanding of this method. (See [124], [95], [90], [158], [73], [157], [117], [137].)

Faddeev [34] proved the spectral theorem for the trivial representation χ. His results were later generalized by Venkov (see [7]). One can also find the description of Faddeev's paper in Lang [121]. We mention a modification of Faddeev's method from the standpoint of non-stationary scattering theory (see Lax and Phillips [128]).

There exist at least four proofs of meromorphic extendability of Eisenstein–Maass series for Fuchsian groups of the first kind. Two of them are accredited to Selberg. The first method, dated at approximately 1950, was described by him in his lectures [163]. The published modified variant of this method can be found in Neunhoffer (1973) [137]. The precise description of the second method is contained in the published report by Selberg to the Mathematical Congress in Stockholm (1962) [165]. The second method is more universal, and, in developing this method, Langlands could prove meromorphic continuation of the Eisenstein series in the most general case of a reductive group (see [124] and [95]). The third method (1965) is accredited to Faddeev [34] and the fourth, final one (1974) is due to Lax and Phillips [128] influenced by Faddeev and Pavlov [35]) (see also de Verdière [64]). These various ways of proving the theorem came about, firstly, because of the absence of any detailed, widely published proof from Selberg himself. Secondly, the assertion on meromorphicity of the function $E(z;s)$ and, especially, on the absence of poles on the line $\mathrm{Re}\, s = \frac{1}{2}$ is very refined and hardly provable. Thus, for a better understanding of the essence of the matter, it is reasonable to have several means of proving these assertions. In this context, it is appropriate to recall the classic example of Gauss's eight proofs of the quadratic reciprocity law in number theory. To illustrate the refinement of the assertion on the absence of poles of the function $E(\,\cdot\,, \frac{1}{2}+ic)$, we say that even in the particular case of modular group Γ it is equivalent to the asymptotic law of distribution of natural primes which was proved by Hadamard and de la Vallée Poussin ($\zeta(s) \neq 0$ on the line $\mathrm{Re}\, s = 1$).

All the above-mentioned methods have a general application. If we restrict ourselves to particular examples of arithmetic groups, to modular groups, for example, then it is known that another proof of meromorphicity exists by means of Poisson's summation formula [90].

Chapter 5.

The most general result in deriving the explicit formulas for Fourier coefficients of analytic automorphic forms and functions is due to Lehner (see [19] and references therein), who for this purpose generalized the classic circle method of Hardy–Ramanujan–Littlewood–Rademacher [94].

Fourier expansions described in this chapter are referred to as expansions at parabolic points. One can also define the Fourier expansions at hyperbolic points and in this case there is no need to assume F to be non-compact. See Fay [78] for this question.

For detailed proof of Theorem 5.2 see [163], [117], [7].

We now comment on Theorems 5.3 and 5.4. In the form stated here, these theorems are due to Fay [78]. The less detailed Fourier expansion for the kernel of the resolvent of the automorphic Laplacian was considered earlier in Faddeev [34] (see also [7]) and in Neunhoffer [137].

The Kuznetsov formula from Theorem 5.6 is convenient for investigating mean values of the Kloosterman sums. Proskurin generalized [31], [30] this formula to the case where Γ is an arbitrary Fuchsian group of the first kind with a fundamental domain having at least one cusp and $S(n, m; c)$ is the Kloosterman sum with the multiplicator χ of an arbitrary weight k. Thus, he obtained the formula

$$\sum_{1\leq c\leq X} \frac{S(n,m;c)}{c} = \mu_1(m,n)X^{\omega_1} + \mu_2(m,n)X^{\omega_2} + \cdots$$

$$\cdots + \mu_l(m,n)X^{\omega_l} + \underset{X\to\infty}{O}(X^{1/3}\ln X), \tag{13.1}$$

$$X \geq 2, \quad \mu_j(m,n) \in \mathbb{C}, \quad \omega_j = \sqrt{1-4\lambda_j}, \quad \lambda_1 \leq \lambda_2 \leq \cdots \leq \lambda_l$$

are those eigenvalues of automorphic Laplacian which lie on the interval $[0, \frac{1}{4})$; the constant in O depends only on m, n, Γ, k, χ.

For $\Gamma = \mathrm{PSL}(2,\mathbb{Z})$, $k = 0$, $\chi = 1$ there are no eigenvalues less than $\frac{1}{4}$, and for the Kloosterman sums there is the estimate of Weil. Then, instead of (13.1) we have Kuznetsov's estimate [17]

$$\sum_{1\leq c\leq X} \frac{S(n,m;c)}{c} \ll X^{1/6}\ln^2 X.$$

This estimate is the first advance in proving Linnik's hypothesis (see [22]).

Note the important survey by Iwaniec 'Non-holomorphic modular forms and their applications', and Stark's paper on Fourier coefficients for parabolic forms of weight zero in [154].

Chapter 6.

The main references to this chapter are: Selberg [163], [164] and [89], [36], [117], [100], [7] (see also the references to Chapter 12). The exposition of the first and second parts of Chapter 6 uses the survey [6]; for the proof of Theorem 6.3 see [7].

We will now comment on the derivation of the Selberg trace formula, methods and published literature on this subject.

From a philosophical point of view, the derivation of the desired trace formula consists of two steps. The first step includes rather lengthy special calculations up to some plausible conjectures. The second step is a justification of those formal computations and, in particular, it proves that certain remainder terms are decreasing. As for the first step, at present we know of only one method due to Selberg for deriving the trace formula which he described in detail in his lectures (see [163]). It is in fact suggested by the Maass–Selberg relation (4.7) and was later independently discovered by other authors (see [71], [117], [36], [39]). As for justification of the Selberg trace formula, i.e. the second step, here there exist several different variants of proving it (see [163], [39], [36], [128]). At least two of them are due to Selberg. The first was described in his lectures [163] and the second was modified by Arthur in [39]. The derivation of the trace formula for some arithmetic groups Γ is discussed also in [71], [39], [109].

Chapter 7.

The chapter is devoted to the development of Selberg's ideas which are contained in [164] and his lectures in Göttingen [163]. We have reason to suppose that the main body of results of this chapter was, to some extent, known to Selberg at the beginning of the 1950's.

The definition of the zeta-function $Z(s;\Gamma;\chi)$ was given by Selberg in [163] (for one-dimensional representation χ). The theory of the Selberg zeta-function for co-compact Fuchsian groups was exposed in detail in the first volume of the monograph by Hejhal [100].

Theorem 7.1 and the results on the location of zeros and poles of $Z(s;\Gamma;\chi)$ are contained in [7]. Theorem 7.2 and the factorization formula for the determinant of an automorphic scattering matrix (7.5) are proved in [10]. The description of a unitary isometry T is taken from Ray and Singer [155]. It should be noted here that the connection of the analytic Ray–Singer torsion with the value of the Selberg zeta-function at the point $s = 1$ was found in [156].

Theorem 7.3 in the context of a co-compact Fuchsian group Γ and one-dimensional representation χ was obtained by Hejhal [100] and Randol [153] and was expanded on by Venkov in the case of general Fuchsian groups of the first kind. This theorem is proved in [7]; in certain special cases it also allows an essential refinement (see Chapter 8, and [7], [9]).

Theorem 7.4 originated, apparently, from Selberg and independently from Huber [106] who considered the case of a co-compact Fuchsian group; see also [153]. This theorem was proved in [7] for Fuchsian groups with non-compact fundamental domain. Sharpening the remainder term in (7.11) is given in [108] for special cases.

We now comment on those results in Selberg's theory of zeta-function that were not considered in Chapter 7. First of all, there are various refinements of the assertions indicated here and so-called Ω-theorems for remainder terms in formulas (7.9), (7.11) which can be obtained in the case of non-compact Fuchsian group Γ. These, and many other more specialized questions, were examined in [100]. There is also a possibility of strengthening the results of this chapter in the case of arithmetic Fuchsian groups; references [94], [5], [19], [28], [108] are devoted to this subject. The reader is also referred to [79].

Chapter 8.

The problems of this chapter originate mainly in the papers by Selberg. Many of the questions discussed here are of independent interest if viewed from the standpoint of general spectral theory.

In exposing the results on eigenvalues of the discrete spectrum of automorphic Laplacians lying on the continuous one, we tend to follow, in the main, monograph [7]. Theorem 8.1 was apparently known to Selberg (judging from the note in his lectures [163]). All the results on infinity of the discrete spectrum for Fuchsian groups with non-trivial commensurators including the groups generated by reflection and commensurable with them (see, in particular, Theorems 8.2 and 8.3) are proved in detail in [7]. The applications of spectral theory of automorphic functions to classical boundary problems in mathematical physics are also considered here and Theorem 8.4 is derived along with other results.

The problems of small eigenvalues were first confirmed by the report by Selberg [166]. The asymptotics for $N_\Gamma(T; z, z')$ obtained by Huber [105] in the case of strictly hyperbolic groups Γ was extended to the general case in Patterson [140]. Selberg's example is discussed in detail by Zograf in his thesis (LOMI, Academy of Sciences, Leningrad, 1982); Theorem 8.5 is proved here as well. Partial use of the results of [57], [63], [51] to the case of general Fuchsian groups of the first kind is in [16] (see also [58]).

Insuperable difficulties which stand in the way of proving Weyl's asymptotic law or even Roelcke's conjecture (or Roelcke–Selberg conjecture, as it is called by certain authors) for general Fuchsian groups of the first kind lead one to the conclusion that these formulas may be false in such a general context. The paper which reinforces these doubts is by de Verdière [65]. Roughly speaking, the author proved that on a Riemannian manifold with cusps and an arbitrary metric, the Laplace operator has no discrete eigenvalues on the continuous spectrum, i.e. their existence means special metric. Thus, there arises the question whether the metric defined by the automorphic Laplacian for a general Fuchsian group of the first kind is specific in this sense. Phillips, Sarnak, Deshouillers and Iwaniec (see [148], [149], [69] and [103]) keeping in mind the results we have indicated, consider it likely that for general Fuchsian groups of the first kind the discrete spectrum of the automorphic Laplacian lying on the continuous one is either very small (i.e. a finite number of eigenvalues) or it is completely absent. This conjecture was first stated in [148], and we refer to it as the Phillips–Sarnak conjecture. An interesting theorem, as well as certain computer calculations on which we will comment later, is a motivation for discussing this conjecture. The theorem is as follows (see [69]). Assuming valid 'almost' standard hypothesis of analytic number theory such as the extended Riemann hypothesis or Lindelöf hypothesis for special Dirichlet series arising from the group $\Gamma_0(q)$, as well as assuming multiplicity of the discrete spectrum for $\mathrm{PSL}(2,\mathbb{Z})$ of the automorphic Laplacian to be equal to 1, the authors bounded from below the order of the integral of φ'/φ in the Weyl–Selberg formula (7.8) as $\gg T^{2-\epsilon}$ for any $\epsilon > 0$, for a general discrete group Γ from the Teichmüller space $T(\Gamma_0(q))$. This is certainly in contrast with the situation of the congruence group Γ. Computer investigations carried out in the case of $\Gamma_1(24)$ (see [149]) show that the zeros of the zeta-function of Selberg are removed from the line $\mathrm{Re}\, s = \frac{1}{2}$ into the left half-plane by a weak deformation in the Teichmüller space (the question is, certainly, on some first zeros).

The situation, then, is very interesting, we have mutually contradicting conjectures, and in order to discover the truth some new information is needed.

We now comment on some interesting problems which are not addressed in the main text of this book. The question is of applications in analytic number theory. There is an interesting problem on mutual relation of the spectral theory of automorphic functions and the theory of Riemann zeta-function. The Selberg trace formula for a modular group allows us, for example, to approximate with multiple precision the ψ-function of Chebyshev by the expression depending only on eigenvalues λ_j of $A(\Gamma_{\mathbb{Z}};1)$ and norms of primitive hyperbolic conjugacy classes in $\Gamma_{\mathbb{Z}}$ (see [5]). Other interesting results were obtained by Iwaniec in estimating mean values of Riemannian $\zeta(s)$ and in the additive divisor problem (see [154]).

In conclusion, we refer to some papers ([60], [101], [102]) that are devoted to computer calculations of eigenvalues for $A(\Gamma_{\mathbb{Z}};1)$.

Chapter 9.

The statement that the spectrum of automorphic Laplacian in the case of a strictly hyperbolic Fuchsian group determines uniquely its genus and the spectrum of lengths of closed geodesics was made by Selberg in [164] and Huber in [106].

Theorem 9.1 is attributed to Gelfand (see [14]). Lemma 9.1 is classical and is attributed to Fricke and Klein (see [111]). We carry out the adapted proof of this lemma using McKean [132].

Theorem 9.2 was proved by McKean in [132]. Inequality (9.5), on which the proof of McKean is based, is known in the literature as the inequality of Mumford. It is obtained in [136]. For the definition and main properties of a Teichmüller space, see [38].

Theorems 9.3 and 9.4 are due to Wolpert [184]. For their proofs, we direct the reader to the original paper [184]. Here we state only that in the 'typical' situation, the Riemann surface can be uniquely determined by the spectrum of lengths of closed geodesics in the spirit of classical ideas of Klein and Fricke [111]. It is interesting to note that in the proof, some spectral information is used, namely, a two-sided estimate for the first eigenvalue of the Laplace operator on a compact Riemann surface in terms of the shortest homologically trivial geodesic chain [162].

In connection with the results of [184] a natural question arises of an effective description of the submanifold $V_g \subset T_g$ mentioned in the statement of Theorem 9.3. Unfortunately, nothing is known about it as yet. As Vignéras indicated in [178] the set V_g is not empty for certain values of g. The corresponding examples were obtained by means of strictly hyperbolic Fuchsian groups connected with various orders of indefinite quaternion algebras over real quadratic fields with the number of ideal classes equal to 2 (for details, see [178]). Thus, the Gelfand conjecture turns out to be invalid, since there exist isospectral but not isometric Riemann surfaces.

We note that in [59] Buser indicated the following fact: in each class of isospectral Riemann surfaces of genus g the number of mutually non-isometric surfaces does not exceed $\exp(507\, g^3)$. We hope that new information on the construction of submanifolds V_g will soon be available.

Notice in conclusion that it should be very interesting to clarify to what extent the spectral measure of the automorphic Laplacian $A(\Gamma;1)$ determines the general Fuchsian group of the first kind, i.e. to use (even partially) the results of this chapter in the case of groups with non-compact fundamental domain of finite measure. Unfortunately, we are not aware if it is in use.

Chapter 10.

The main references to this chapter are: Heath-Brown and Patterson [96], and Kubota [113], [115]. As for the history of the Kummer problem, as well as some interesting facts on Gauss sums, see the survey by Berndt and Evans [50].

A various number of proofs of the quadratic reciprocity law are known of. One of these is due to Cauchy, and it uses, instead of (10.4), the following transformation formula

$$\theta(-\tfrac{1}{z}) = \sqrt{-iz}\,\theta(z)$$

for the classical theta-function

$$\theta(z) = \sum_{n\in\mathbb{Z}} e^{2\pi i n^2 z}, \quad z \in \mathbb{C}, \quad \operatorname{Im} z > 0.$$

It seems appropriate here to note that the theta-function is one of the first known automorphic forms.

In a more general sense, for primes p and q of a field $\mathbb{F}$ of algebraic numbers containing the roots of power k from 1 a residue symbol $(p/q)_k$ of order k is defined.

The reciprocity law expresses in the explicit form the ratio $(p/q)_k(q/p)_k^{-1}$ by means of the numbers p and q. In the case where $\mathbb{F}$ is a cyclotomic field of degree k, k is a prime, the reciprocity law was proved by Kummer and applied to investigating the Fermat equation. The explicit form of general reciprocity law was obtained by Shafarevitch (information on reciprocity laws and a useful bibliography can be found in Faddeev [33]). These proofs do not use the Gauss sums. The fact is that in a general case, one cannot obtain the explicit formulas of the type of (10.4). There are no automorphic functions of the type of theta-functions which allow one to generalize the Cauchy proof.

Note also the distinguished paper by Patterson [142] in which the author generalized Theorem 10.1 on the Gauss sums of an arbitrary order. We also refer to some new papers by the same authors in connection with these problems (see [143–145]).

Chapter 11.

The main references to this chapter are: Selberg [164], [165], [167], Gelfand [14], Langlands [124], Harish–Chandra [95], Osborne and Warner [139]. As an introduction, it is useful to read the report by Terras in [103].

The reader can find the elements of the theory of groups and Lie algebras in Serre [169].

The significance of the representation theory of Lie groups in the theory of automorphic functions was first clearly understood in Gelfand, Graev, Pjatetskiĭ-Shapiro and Godement (see, for example, [14], [15], [88], [89]).

As for 'the computation' of orbital integrals, see [180], [139] and bibliography therein.

We refer to one important application of the Selberg trace formula, namely, the calculation of multiplicities $m(U, \Gamma)$ of irreducible representations entering the regular representation $R_{\Gamma\backslash G}$ (see (12.10)). In such a way one obtains, in particular, the formulas for dimensions of the spaces of analytic automorphic forms (see [165], [122], [180], [139] and bibliography in this book).

Applications to arithmetic of number fields of spectral theory of automorphic functions in hyperbolic spaces can be found in the interesting papers by Elstrodt, Mennicke and Grünewald [75], [76].

In [70] Donnelly examines the cuspidal spectrum of an automorphic Laplacian on general symmetric space relative to arithmetic discrete group. The author obtained a certain estimate from above for the corresponding distribution function of eigenvalues $N(\lambda)$ which is a considerable refinement of the results of Borel [52] and Garland [83]. We note that at present, there are no reasonable estimates from below for these functions in such a general situation.

We also refer to [133] in which Morris, developing the ideas of Harder [93] and Langlands [124], constructed the theory of Eisenstein series on an arbitrary reductive group defined over a global functional field.

The ideas given in this chapter are a simplified outline of an extensive, incomplete theory which is difficult to understand. It is no wonder that even specialists have at present some ambiguity in interpreting viewpoints in the theory. (For example, see Chapter 7 in [124]. The corresponding discussion is in [139], [127]). The derivation of the Selberg trace formula and decomposition of the representation $R_{\Gamma\backslash G}$ onto irreducible ones in the special case of the group G of split rank 1 (the question is on Γ with non-compact quotient $\Gamma\backslash G$ of finite volume) are the clearest aspects of the theory. It is well-recognized (see [167], [84]) that such G has only the Γ of rank 1, i.e. $\Gamma\backslash G$ has all cusps of type one. If we eliminate non-arithmetic Γ which turn out to exist here (see [23], [11]), then deriving the Selberg trace formula and examing the regular representation $R_{\Gamma\backslash G}$ are not difficult, and the situation is much the same as for GL(2) (see [2], [39], [120], [181]). Next in degree of difficulty is the case that is close to the former, and it also may be considered as a clear one. The question is justification of the trace formula for the group Γ of rank 1 lying in arbitrary reductive G. (If the rank of G is greater than 1 then it is unnecessary to lay the arithmeticity condition upon in order to simplify the derivation since this condition is fulfilled automatically by Margulis theorem, see [24], [25], [26], [134]). The Selberg trace formula is obtained here by Warner and Osborne (see [138]), and this is the most general example of the pair of G and Γ for which the Selberg trace formula exists in such a form as it was published by Selberg himself in [164] for Fuchsian groups of the first kind with non-compact fundamental domain. In this context, one should mention the paper by Efrat [72] in which deriving the Selberg trace formula is carried out on the product of hyperbolic planes. (See also [186]).

Concluding this commentary, we refer to the outstanding series of comparatively new works by Arthur which are devoted to the general Selberg trace formula. The best reference containing the main ideas of the derivation of the Selberg–Arthur trace formula is the report [45]. (The report by Labesse [119] which also contains the applications of the trace formula is a useful supplement to this subject). The paper [46] is realized in more classical terms and is devoted to the trace formula for the group $SL(n,\mathbb{Z})$ acting on the corresponding symmetric space of symmetric positive definite matrices. Here the main spectral difficulties are illustrated which are in the way of deriving the traditional Selberg trace formula in this situation. Finally, in [47], [48] following the ideas of [45], further details and interpreting individual summands of general trace formula are continued.

Chapter 12.

The survey by Panchishkin [27] may turn out to be useful. He comments on papers devoted to the connection between automorphic functions, representations and L-functions; also included is a large bibliography which complements the one presented in this book.

The first step towards proving the Langlands conjecture was made by Langlands himself. He considered two-dimensional representations (see (12.6)). It is acknowledged that the image relative to σ of the Galois group $\mathrm{Gal}(\mathbb{F}'/\mathbb{F})$ in $\mathrm{PGL}_2(\mathbb{C})$ is isomorphic to one of the following groups:

(1) to dihedral group (σ is the monomial representation);
(2) to the group A_4 (σ is the tetrahedral representation);
(3) to the group S_4 (σ is the octahedral representation);
(4) to the group A_5 (σ is the icosahedral representation).

Langlands proved the conjecture on coincidence of L-functions and, therefore, the Artin conjecture in cases (2) and (3) (see [85]). We are not able to discuss here, even briefly, the proof of Langlands. We restrict ourselves instead to a statement of the so-called functoriality principle used in this proof, which is very significant to many other questions. The principle is accredited to Langlands.

Let G be a connected reductive algebraic group over a global (or local) field $\mathbb{F}$ and ${}^LG$ be a dual Langlands group. In order to construct the group ${}^LG$, one must determine its connected component ${}^LG^0$; it is a reductive group over $\mathbb{C}$, root data of which are obtained by inversion of the root data of the group G. Then ${}^LG$ can be constructed as a semi-direct product of ${}^LG^0$ by the Galois group $\mathrm{Gal}(\mathbb{F}'/\mathbb{F})$ of a certain extension $\mathbb{F}'$ of the field $\mathbb{F}$ (over which G splits), so we have the following exact sequence:

$$1 \to {}^LG^0 \to {}^LG \xrightarrow{P_1} \mathrm{Gal}(\mathbb{F}'/\mathbb{F}) \to 1.$$

(See the precise definition in [54]; we only note, as an example, that for $G = \mathrm{GL}_n$ we have ${}^LG^0 = \mathrm{GL}_n$, and, for $G = \mathrm{SP}_{2n}$ ${}^LG^0 = \mathrm{SO}_{2n+1}$.) Let us now be given another (except G) connected reductive algebraic group H over $\mathbb{F}$, and let

$$1 \to {}^LH^0 \to {}^LH \xrightarrow{P_2} \mathrm{Gal}(\mathbb{F}''/\mathbb{F}) \to 1$$

be a corresponding exact sequence. We call a continuous homomorphism $u : {}^LH \to {}^LG$ L-homomorphism, if the following conditions are fulfilled:

(1) the restriction of u to ${}^LH^0(\mathbb{C})$ is a complex-analytic homomorphism into ${}^LG^0(\mathbb{C})$;

(2) $\mathbb{F}' \subset \mathbb{F}''$;

(3) the diagram

$$\begin{array}{ccc} & {}^LH \xrightarrow{u} {}^LG & \\ P_2 \swarrow & & \searrow P_1 \\ \mathrm{Gal}(\mathbb{F}''/\mathbb{F}) & \to & \mathrm{Gal}(\mathbb{F}'/\mathbb{F}) \end{array}$$

is commutative.

Now let $\mathbb{F}$ be a global field, let $u : {}^LH \to {}^LG$ be a L-homomorphism, let π be an irreducible admissible representation of the group $H(\mathbb{A}_{\mathbb{F}})$ and let $\pi = \otimes_v \pi_v$ be a factorization of π into the restricted tensor product (over all the points v of the field $\mathbb{F}$) of the representations π_v of the groups $H(\mathbb{F}_v)$. For almost all v there corresponds (by common rule, as in (12.18)) to the representation π_v the conjugacy class of a semi-simple element h_v in the Langlands group ${}^LH(\mathbb{C})$. It is supposed (and this is the functoriality principle) that under the mentioned conditions there exists an irreducible admissible representation $\pi' = \otimes_v \pi'_v$ of the group $G(\mathbb{A}_{\mathbb{F}})$ such that for almost all v the class h'_v corresponding to π'_v is equal to $u(h_v)$. (For a more detailed and precise description of the functoriality principle, see [54], [125]). The functoriality principle contains as a special case the so-called problem of base change. The problem is to connect with each automorphic representation of the group $G(A_{\mathbb{F}})$ an automorphic representation of the group $G(A_{\mathbb{F}'})$; here G and $\mathbb{F}$ are as above, and $\mathbb{F}'$ is a finite Galois extension of the field $\mathbb{F}$. The problem of base change was solved by Langlands, who made use of the earlier investigations of Saito and Shintani, for $G = \mathrm{GL}_2$ in the case where the extension $\mathbb{F}'/\mathbb{F}$ is solvable. The solution of this problem was the first main step in Langlands' proof of the Artin conjecture for two-dimensional representations.

Without pretending to any completeness of the bibliography in this respect, we direct the reader to the extensive paper by Arthur and Closel [49] devoted to the problem of base change for $\mathrm{GL}(n)$.

References

1. Hadamard J., Non-Euclidean geometry in the theory of automorphic functions, GITTL, Moscow, 1951 (Russian).
2. Venkov A. B., Expansion in automorphic eigenfunctions of the Laplace–Beltrami operator in classical symmetric spaces of rank one, and the Selberg trace formula, *Trudy Mat. Inst. Steklov* **125** (1973), 6–55 (Russian); English transl. in *Proc. Steklov Inst. Mat.* **125** (1973).
3. Venkov A. B., On the Selberg trace formula for SL(3, $\mathbb{Z}$), *Dokl. Akad. Nauk SSSR*, **228** (1976), N 2, 273–276 (Russian).
4. Venkov A. B., On the Selberg trace formula for SL(3, $\mathbb{Z}$), *Zap. Nauchn. Sem. Leningrad. Otdel. Mat. Inst. Steklov* (LOMI), **63** (1976), 8–66 (Russian); English transl. in *J. Soviet Math.* (1979), v. **12**, 384–424.
5. Venkov A. B., A formula for the Chebyshev psi-function, *Mat. Zametki*, **23** (1978), 497–503 (Russian); English transl. in *Math. Notes* **23** (1978).
6. Venkov A. B., Spectral theory of automorphic functions, the Selberg zeta-function, and some problems of analytic number theory and mathematical physics, *Uspehi Mat. Nauk* **34** (1979), N 3 (207), 69–135 (Russian); English transl. in *Russian Math. Surveys* **34** (1979).
7. Venkov A. B., Spectral theory of automorphic functions, *Trudy Mat. Inst. Steklov*, **153** (1981), 3–171 (Russian); English transl. in *Proc. Steklov Inst. Math.* (1982), issue **4**, 1–163.
8. Venkov A. B., Dirichlet-series associated with defining equations and continued fractions in the theory of automorphic functions, *Trudy Mat. Inst. Steklov*, **158** (1981), 31–44 (Russian).
9. Venkov A. B., On automorphic scattering matrix for Hecke group $\Gamma(2 \cos \pi/q)$, *Zap. Nauchn. Sem. Leningrad. Otdel. Mat. Inst. Steklov*, **109** (1981), 34–40 (Russian).
10. Venkov A. B. and Zograf P. G., On analogs of Artin factorization formulas in spectral theory of automorphic functions related to induced representations of Fuchsian groups, *Izv. Akad. Nauk SSSR, Ser. Mat.* **46** (1982), N 6, 1150–1158 (Russian).
11. Vinberg E. B., Discrete groups generated by reflections in Lobachevsky spaces, *Mat. Sb.* **72** (114), 1967 (Russian).
12. Vinogradov A. I., and Takhtazhan L. A., The Gauss–Hasse hypothesis on real quadratic fields with class number one, *Grelle Math. J.*, **335** (1982), 40–86.
13. Vinogradov I. M., Selected works, Moscow (1952) (Russian).
14. Gelfand I. M., Automorphic functions and the theory of representations, Proc. ICM Stockholm (1962), Uppsala, Almqvist–Wiksells (1963).
15. Gelfand I. M. et al., Generalized functions, N 6, Moscow, Nauka, 1966 (Russian).
16. Zograf P. G., Fuchsian groups and small eigenvalues of the Laplace operator, *Zap. Nauchn. Sem. Leningrad. Otdel. Mat. Inst. Steklov* (LOMI), **122**, 24–29 (Russian).
17. Zograf P. G., On spectrum of automorphic Laplacians in spaces of parabolic functions, *Dokl. Akad. Nauk SSSR*, **269** (1983), N **4**, 802–805 (Russian).
18. Kuznetsov N. V., Petersson conjecture for forms of weight zero and the Linnik conjecture, Preprint, Khabarovsk (1977) (Russian).
19. Kuznetsov N. V., Petersson conjecture for cusp forms of weight zero and Linnik conjecture. Sums of Kloosterman sums. *Mat. Sb.* **111** (**153**) (1980), 334–383; English transl. in *Math. USSR Sb.* **39** (1981).
20. Kuznetsov N. V., Convolution of Fourier coefficients of Eisenstein–Maass series, *Zap. Nauchn. Sem. Leningrad. Otdel Mat. Inst. Steklov* (LOMI) **129** (1983), 43–84 (Russian).
21. Kubilyus I. P., On certain problems of geometry of primes, *Mat. Sb.* **31** (**73**) (1952), 507–542 (Russian).
22. Linnik Ju. V., Additive problems and eigenvalues of the modular operators, in Proc. ICM Stockholm (1962), Uppsala, Almqvist–Wiksells (1963).

23. Makarov V. S., On one class of discrete groups of the Lobachevsky space which have an infinite fundamental domain of finite measure, *Dokl. Akad. Nauk SSSR*, **167** (1966), N 1, 30–33 (Russian).
24. Margulis G. A., Arithmetical properties of discrete subgroups, *Uspehi Mat. Nauk*, **29** (1974), 49–98 (Russian).
25. Margulis G. A., Discrete groups of motions of manifolds of non-positive curvature, *Proc. Internat. Congr. of Math., Vancouver*, **2** (1974), 21–34.
26. Margulis G. A., Arithmeticity of irreducible lattices in semisimple groups of a rank greater than 1, Addition to Russian translation of the book Raghunathan M., Discrete subgroups of Lie groups, Moscow (1977), 277–313 (Russian).
27. Panchishkin A. A., Modular forms. Itogi nauk and tekhn. *Algebra, Topology, Geometry*, **19** (1981), 135–180 (Russian).
28. Proskurin N. V., On Linnik conjecture, *Zap. Nauchn. Sem. Leningrad. Otdel. Mat. Inst. Steklov* (LOMI), **91** (1979), 94–118 (Russian).
29. Proskurin N. V., An estimate of eigenvalues of Hecke operators in the space of cusp forms of weight 0, *Zap. Nauchn. Sem. Leningrad. Otdel. Mat. Inst. Steklov*, (LOMI), **82** (1979), 136–143 (Russian).
30. Proskurin N. V., Summation formulas for general Kloosterman sums, *Zap. Nauchn. Sem. Leningrad Otdel. Mat. Inst. Steklov* (LOMI), **82** (1979), 103–135 (Russian).
31. Proskurin N. V., On general Kloosterman sums, Preprint LOMI, P-3-80 (1980), 1–36 (Russian).
32. Tikhonov A. N. and Samarskii A. A., Equations of mathematical physics, Moscow (1953) (Russian).
33. Faddeev D. K., On 9-th problem of Hilbert; in the book: Problems of Hilbert, Moscow, Nauka (1969), 131–140 (Russian).
34. Faddeev L. D., Expansion in eigenfunctions of the Laplace operator on the fundamental domain of a discrete group on the Lobachevsky plane, *Trudy Moscov. Mat. Obsc.*, **17** (1967), 323–350; English transl. in *Trans. Moscow Math. Soc.* **17** (1967).
35. Faddeev L. D. and Pavlov B. S., Scattering theory and automorphic functions, *Zap. Nauchn. Sem. Leningrad. Otdel. Mat. Inst. Steklov* (LOMI), **27** (1972), 161–193; English transl. in *J. Soviet Math.* **3** (1975) N 4.
36. Faddeev L. D., Venkov A. B. and Kalinin V. L., Non-arithmetical derivation of the Selberg trace formula, *Zap. Nauchn. Sem. Leningrad Otdel. Mat. Inst. Steklov* (LOMI), **37** (1973), 5–42; English transl. in *J. Soviet Math.* **8** (1977), N 2.
37. Fomenko O. M., Applications of modular form theory to number theory. Itogi nauki i tekhn. *Algebra, Topology, Geometry* **15** (1977), 5–91 (Russian).
38. Ahlfors L., Lectures on Quasiconformal mappings, Van Nostrand, 1966.
39. Arthur J., The Selberg trace formula for groups of F-rank one. *Ann. of Math.* (1974), v. **100**: 2.
40. Arthur J., A trace formula for reductive groups I: terms associated to classes in $G(\mathbb{Q})$, *Duke Math. J.* (1978), v. **45**, 911–952.
41. Arthur J., Eisenstein series and the trace formula, *Proc. Symp. Pure Math. AMS*, Part 1 (1979), v. **33**, 253–274.
42. Arthur J., A trace formula for reductive groups. II : applications of a truncation operator, *Comp. Math.* (1980), v. **40**, 87–121.
43. Arthur J., The trace formula in invariant form, *Ann. of Math.* (1981), v. **114**, 1–74.
44. Arthur J., On a family of distributions obtained from Eisenstein series I: Application of the Paley–Wiener theorem II Explicit formulas, *Amer. J. of Math.* (1982), v. **104**, N 6, 1243–1336.
45. Arthur J., The trace formula for reductive groups, *Publ. Math. L'Université Paris VII* (1983), v. **15**, 1–42.
46. Arthur J., The trace formula for noncompact quotient, *Proc. Intern. Congr. Math., Warsaw* (1984), v. **2**, 849–860.
47. Arthur J, A measure on the unipotent variety, *Can. J. Math.* (1985), v. **37**, N 6, 1237–1274.
48. Arthur J., On a family of distributions obtained from orbits, *Can. J. Math.* (1986), v. **38**, N 1, 179–214.
49. Arthur J., Closel L., Base change for GL(n), Preprint of University of Toronto (1986).
50. Berndt B. C., Evans R. J., The determination of Gauss sums, *Bull. AMS* (1981), v. **5**, N 2, 107–129.
51. Besson G., Sur la multiplicité de la premiére valeur propre 'des surfaces riemanniènnes, *Ann.*

Inst. Fourier (1980), v. **30**, N 1, 109–128.
52. Borel A., Cohomology of arithmetic groups, *Proc. Internat. Congr. Math. Vancouver* (1974), v. **1**, 435–442.
53. Borel A., Formes automorphes et séries de Dirichlet (d'après R.P. Langlands) *Lect. Notes Math.* (1976), v. **514**, 183–222.
54. Borel A., Automorphic L-functions, in [49] P. **2**, 27–61.
55. Borel A., Casselman W. (eds.), Automorphic forms, representations and L-functions, *Proc. Symp. Pure Math., AMS* (1979), v. **33**, part 1, 2.
56. Bruggeman R. W., Fourier coefficients of cusp forms, *Inventione Math.* (1978), v. **45**, 1–18.
57. Buser P., Riemannsche Flächen mit Eigenwerten in $(0, 1/4)$, *Comm. Math. Helv.* (1977), Vol. **52**, N 1, 25–34.
58. Buser P., On Cheeger inequality $\lambda_1 \geq h^2/4$, [11], 29–38.
59. Buser P., Sur le spectre de longueurs des surfaces de Riemann, *C.R. Acad. Sci. Paris*, Ser. I (1981), t. **292**, 487–489.
60. Cartier P., Some numerical computations relating to automorphic functions. Computers in number theory, Acad. Press (1977), 37–48.
61. Chandrasckharan K., Arithmetical functions, Berlin, Springer (1970).
62. Chen Ying-Run, The lattice points in a circle, *Chinese Math.* (1963), v. **4** :2, 322–339.
63. Cheng, S.-Y., Eigenfunctions and model sets, *Comm. Math. Helv.* (1976), v. **51**, N 1, 43–55.
64. Colin de Verdière Y., Une nouvelle demonstration du prolongement méromorphe de séries d'Eisenstein, *C.R. Acad. Sci. Paris*, Série I, t. **293**, 361–363.
65. Colin de Verdière Y., Pseudo-Laplacians I and II, *Ann. de Inst. Fourier* (1983), v. **32**, N 3, v. **33**, N 3, 87–113.
66. Deligne P., La conjecture de Weil, *Publications Math.* (1974), N 43, 279–307.
67. Delsarte J., Le gitter fuchsien. Oeuvres de Jean Delsarte, *Tome* II, 829–845, Paris (1971).
68. Deshouillers J. M. and Iwaniec H., Kloosterman sums and Fourier coefficients of cusp forms, Preprint, Université Bordeaux 1 (1982).
69. Deshouillers J. M., Iwaniec H., Phillips R. S. and Sarnak P., Maass cusp forms, *Proc. Natl. Acad. Sci. USA* (1985), v. **82**, 3533–3534.
70. Donnelly H., On the curpidal spectrum for finite volume symmetric spaces, *J. Differential Geometry* (1982), v. **17**, 239–253.
71. Duflo M., Labesse J.-P., Sur la formule des traces de Selberg, *Ann. Scient. Ec. Norm. Sup., 4-e series* (1971), v. **4**, 193–284.
72. Efrat I. Y., The Selberg trace formula for $\mathrm{PSL}(2, \mathbf{R})^n$. *Memoirs of AMS* (1987), v. **65**, N 359, 1–111.
73. Elstrodt J., Die Resolvente sum Eigenwertproblem der automorphen Formen in der hyperbolischem Ebene, *Math. Ann.* (1973), v. **203**, 295-330; *Math. Zeit.* (1973), v. **132**, 99–134; *Math. Ann.* (1974), v. **208**, 99–132.
74. Elstrodt J., Die Selbergsche Spurformel für kompakte Riemannsche Flächen, Hauptvortrag auf der Jahrestagung der Deutschen Mathematiker-Vereinigung in Hamburg 1979, Preprint Univ. Munster (1980).
75. Elstrodt J., Grünewald F. and Mennicke J., Eisenstein series on three-dimensional hyperbolic space and imaginary quadratic number fields, *J. Reine Angew. Math.* (1985), v. **360**, 160–213.
76. Elstrodt J., Grünewald F. and Mennicke J., Arithmetic applications of the hyperbolic lattice point theorem, Preprint University Munster (1987).
77. Erdös P. and Turan P., On a problem in the theory of uniform distribution. I, II. *Indagationes* (1948), v. **10**, 307–378, 406–413.
78. Fay J. D., Fourier coefficients of the resolvent for a Fuchsian group, *J. Reine Angew. Math.* (1977), v. **293/294**, 143–203.
79. Fischer J., An approach to the Selberg trace formula via the Selberg zeta-function, *Lecture Notes in Math.*, Springer, v. **1252** (1987).
80. Flath D., Decomposition of representations into tensor products, in [49].
81. Fricke R., Automorphe Funktionen mit Einschluss der Elliptischen Modulfunktionen, *Encyklopädie der mathematischen Wissenschaften*, BII4, Leipzig, Teubner (1913), 349–470.
82. Fried D., The zeta-functions of Ruelle and Selberg 1, *Ann. Sci. Ec. Norm. Sup.*, 4 serie (1986), v. **19**, 491–517.
83. Garland H., The spectrum of noncompact G/Γ and the cohomology of arithmetic groups, *Bull. Amer. Soc.* (1969), v. **75**, 807–811.
84. Garland H. and Raghunathan M. S., Fundamental domains for $\mathbf{R}$-rank 1 semi-simple Lie

groups, *Ann. of Math.* (1970), v. **92**, 279–326.
85. Gelbart S., Automorphic forms and Artin's conjecture, *Lect. Notes Math.* (1977), v. **627**, 241–276.
86. Gérardin P. and Labesse J. P., The solution of a base change problem for GL(2) (following Langlands, Saito, Shintani) in [49], P. 2, 115–133.
87. Geometry of the Laplace operator, *Proc. Symp. Pure Math. AMS*, v. **36**, Providence (1980).
88. Godement R., Introduction aux travaux de A. Selberg, Paris, Sem. Bourbaki, exp. **144** (1957).
89. Godement R., La formule des traces de Selberg, Paris, Sem. Bourbaki, exp. **244** (1962).
90. Godement R., The decomposition of $L^2(G/\Gamma)$ for $\Gamma = \mathrm{SL}(2, \mathbb{Z})$. *Proc. Symp. Pure Math., AMS* (1966), v. **9**, 211–224.
91. Godement R., The spectral decomposition of cusp-forms. *Proc. Symp. Pure Math. AMS* (1966), v. **9**, 110–127.
92. Godement R., Jacquet H., Zeta functions of simple algebras, *Lect. Notes Math.* (1972), v. **260**.
93. Harder G., Chevalley groups over function fields and automorphic forms, *Ann. of Math.* (1974), v. **100**, 249–306.
94. Hardy G. H., Ramanjan S., Asymptotic formulae in combinatory analysis, *Proc. London Math. Soc.* (1918), 75–115.
95. Harish-Chandra, Automorphic forms on semisimple Lie groups, *Lect. Notes in Math.*, v. **62**, Springer (1968).
96. Heath-Brown D. R. and Patterson S. J., The distributions of Kummer sums at prime arguments, *J. Reine Angew. Math.* (1979), v. **31**, 111–130.
97. Hecke E., Mathematische Werke, Göttingen, Vandenhoeck und Ruprecht (1959).
98. Hejhal D. A., Monodromy groups and linearly polymorphic functions, *Acta Math.* (1975), v. **135**, N 1–2, 1–55.
99. Hejhal D. A., The Selberg trace formula and the Riemann zeta-function, *Duke Math. J.* (1976), v. **43**: 3, 441–482.
100. Hejhal D. A., The Selberg trace formula for $\mathrm{PSL}(2, \mathbb{R})$, *Lect. Notes in Math.*, Springer, v. **548** (1976) and v. **1001** (1983).
101. Hejhal D. A., Some observations concerning eigenvalues of the Laplacian and Dirichlet L-series, in *Recent progress in analytic number theory*, v. **2**, Academic Press (1981), 95–110.
102. Hejhal D. A. and Berg B., Some new results concerning eigenvalues of the non-euclidean Laplacian for $\mathrm{PSL}(2, \mathbb{Z})$, Preprint, University of Minnesota, USA (1983).
103. Hejhal D. A., Sarnak P. and Terras A. A. (eds.), The Selberg trace formula and related topics, *Contemp. Math.* (1986), v. **53**.
104. Hilbert D., Mathematische Probleme, *Nachr. Ges. Wisc. Göttingen* (1900), 253–297.
105. Huber H., Über eine neue Klasse automorpher Funktionen und im Gitterpunktproblem in der hyperbolischen Ebene, *Comm. Math. Helv.* (1956), Bd. **30**, 20–62.
106. Huber H., Zur analytischen Theorie hyperbolischer Raumformen und Bewegungsgruppen, *Math. Ann.* (1961), v. **142** , 385–398 (1961), v. **143**, 463–464.
107. Iwaniec H., Mean values for Fourier coefficients of cusp forms and sums of Kloosterman sums, Journées Arithmétiques (1980), 306–321.
108. Iwaniec H., Prime geodesic theorem, *J. Reine Angew. Math.* (1984), v. **349**, 136–159.
109. Jacquet H. and Langlands R. P., Automorphic forms on GL(2), *Lecture Notes in Math.*, v. **114**, Springer (1970).
110. Kazhdan D. A. and Patterson S. J., Metaplectic forms, *Publ. Math. IHES* (1984), v. **59**, 35–142.
111. Klein F. and Fricke R., Vorlesungen über die Theorie der Elliptischen Modulfunktionen/ Automorphenfunktionen, Leipzig, G. Teubner (1896), (1912).
112. Kubota T., Ein arithmetischer Satz über eine Matrizengruppe, *J. Reine Angew. Math.* (1966), v. **222**, 55–57.
113. Kubota T., On automorphic functions and the reciprocity law in number fields. Lecture in Math., v. **2**, Kyoto University (1969).
114. Kubota T., Some results concerning reciprocity and functional analysis, *Actes Congress Intern. Math.*, Nice (1970), t. **1**, 395–399.
115. Kubota T., Two kinds of special functions related to the reciprocity law (1969), Proceedings of the summer institute on number theory, Strong Brook, New York.
116. Kubota T., Some results concerning reciprocity law and real analytic automorphic functions,

Proc. Symp. Pure Math. (1971), v. **20**, 382–395.
117. Kubota T., Elementary theory of Eisenstein series, Tokyo, New York, Kodausha (1973).
118. Kummer E. E., De residuis cubicis disquisitiones nonnullae analyticae, *J. Reine Angew. Math.* (1846), v. **32**, 341–359 (or Coll. Papers v. **1**, 145–163, Berlin, Heidelberg, New York (1975)).
119. Labesse J.-P., La formule des traces d'Arthur–Selberg, *Sem. Bourbaki* (1984–85), N 636, 73–88.
120. Lachaud G., Spectral analysis of automorphic forms of rank 1 groups by perturbation methods, *Proc. Symp. Pure Math., AMS* (1973), v. **26**, 441–450.
121. Lang S., SL(2, $\mathbb{R}$), Addison-Wesley (1975).
122. Langlands R. P., Dimension of spaces of automorphic forms, *Proc. Symp. Pure Math., AMS* (1966), v. **9**, 253–257.
123. Langlands R. P., Eisenstein series, *Proc. Symp. Pure Math., AMS* (1966), v. **9**, 235–252.
124. Langlands R. P., On the functional equations satisfied by Eisenstein series, *Lecture Notes in Math.*, v. **544**, Springer (1976).
125. Langlands R. P., *L*-functions and automorphic representations, Proc. ICM, Helsinki (1978).
126. Langlands R. P., Les debuts d'une formule des traces stable, *Publ. Math. Univ. Paris, VII* (1983), v. **13**.
127. Langlands R. P., Review of the book 'The theory of Eisenstein systems' by S. Osborne and G. Warner, *Bull. AMS* (1983), v. **9**, N 3, 351–360.
128. Lax P. D., Phillips R. S., Scattering theory for automorphic functions, *Ann. Math. Studies*, v. **87**, Princeton, New Jersey (1976).
129. Lehner J., Discontinuous groups and automorphic functions, *AMS*, **8**, Providence (1964).
130. Maass H., Über eine neue Art von nichtanalytischen automorphen Funktionen und die Bestimmung Dirichlet'sher Reihen durch Funktionalgleichungen, *Math. Ann.* (1949), v. **121**: 2, 141–183.
131. Maass H., Lectures on modular functions of one complex variable, Tata Inst. Lecture Notes, Bombay (1964).
132. McKean H. P., Selberg's trace formula as applied to a compact Riemann surface, *Comm. Pure Appl. Math.* (1972), v. **25**, 225–246.
133. Morris L. E., Eisenstein series for reductive groups over global function fields 1, The cusp form case, *Can. J. Math.* (1982), v. **34**, N 1, 91–168.
134. Mostow G. D., Discrete subgroups of Lie groups, *Queen's Papers Pure Appl. Math.* (1978), v. **48**, 65–153.
135. Müller W., Signature defects of cusps of Hilbert modular varieties and values of *L*-series at $s = 1$, *J. Diff. Geom.* (1984), v. **20**, 55–119.
136. Mumford D., A remark on Mahler's compactness theorem, *Proc. AMS* (1971), v. **28**, 289–294.
137. Neunhoffer H., Über die analytische Fortsetzung von Poincaréreihen, Sitz. Heidel. Acad. Wiss. Math.-Natur. Klasse 2 (1973).
138. Osborne M. S., Warner G., The Selberg trace formula I: Γ-rank one lattices, *J. Reine Angew. Math.* (1981), Bd. **324**, 1–113.
139. Osborne M. S., Warner G., The theory of Eisenstein systems, Academic Press (1981).
140. Patterson S. J., A lattice-point problem in hyperbolic space, *Mathematika* (1975), v. **22**, 81–88; (1976), v. **23**, 227.
141. Patterson S. J., A cubic analogue of the theta series. I, II, *J. Reine Angew. Math.* (1977), v. **296**, 125–161, 217–220.
142. Patterson S. J., The distribution of general Gauss sums at prime arguments, In the book: Recent progress in analytic number theory, Edited by H. Halberstam and C. Hooley, v. **2**, Academic Press (1981), 171–182.
143. Patterson S. J., The distribution of certain special values of the cubic Legendre symbol, *Glasgow Math. J.* (1985), v. **27**, 165–184.
144. Patterson S. J., Metaplectic forms and Gauss sums, 1, 2, Preprints Univ. Göttingen (1985–1986).
145. Patterson S. J., The distribution of general Gauss sums and similar arithmetic functions at prime arguments, Preprint Univ. Göttingen (1985).
146. Peterson H., Über einen einfachen Typus von Untergruppen der Modulgruppe, *Archiv der Math.* (1953), Bd. **4**, 308–315.
147. Peterson H., Über die Konstruktion zykloider Kongruenzgruppen in der rationalen Modulgruppe, *J. reine angew. Math.* (1971), Bd. **250**, 182–212.
148. Phillips R. S. and Sarnak P., The Weyl theorem and the deformation of discrete groups,

Comm. Pure Appl. Math. (1985), v. **38**, 853–866.
149. Phillips R. S. and Sarnak P., On cusp forms for co-finite subgroups of PSL(2, **R**), *Invent. Math.* (1985), v. **80**, 339–364.
150. Poincaré H., Sur les groupes des équations linéaires, *Acta Math.* (1884), T. **4**, 201–312.
151. Poincaré H., Mémoire sur les functions zetafuchsiennes, *Acta Math.* (1884), T. **5**, 209–278.
152. Randol B., Small eigenvalues of the Laplace operator on compact Riemann surfaces, *Bull. AMS* (1974), v. **80**, 991–1000.
153. Randol B., The Riemann hypothesis for Selberg's zeta-function and the asymptotic behaviour of eigenvalues of the Laplace operator, *Trans. AMS* (1978), v. **236**: 513, 209–224.
154. Rankin R. A. (ed.), Modular forms, London, Chichester (1984).
155. Ray D. B. and Singer I. M., R-torsion and the Laplacian on Riemannian manifolds, *Advances in Math.* (1971), v. **7**, 145–210.
156. Ray D. B. and Singer I. M., Analytic torsion for complex manifolds, *Ann. of Math.* (1973), v. **98**: 2, 154-177.
157. Roëlcke W., Über die Wellengleichung bei Grenzkreisgruppen erster Art, Sitz. Heidel. Akad. Wiss. Math.-Natur. Klasse (1953/55, 1956), v. **4**.
158. Roëlcke W., Das Eigenwertproblem der automorphen Formen in der hyperbolischen Ebene, *Math. Ann.* (1966), v. **167**, 292–337 (1967), v. **168**, 261–324.
159. Ryavec C., A non-linear interpolation formula, *J. Math. Anal. Appl.* (1982), v. **87**, 468–473.
160. Satake I., Spherical functions and Ramanujan conjecture, *Proc. Sympos. Pure Math., AMS* (1966), v. **9**, 258–264.
161. Schmidt A. L., Minimum of quadratic forms with respect to Fuchsian groups 1, *Gelle J. Math.* (1976), v. **286/287**, 341–368.
162. Schoen R., Wolpert S. and Yau S.–T., Geometric bounds on the law eigenvalues of a compact surface, in **82**, 279–285.
163. Selberg A., Harmonic analysis, 2. Teil, Vorlesungsniederschrift Göttingen (1954).
164. Selberg A., Harmonic analysis and discontinuous groups in weakly symmetric Riemannian spaces with applications to Dirichlet series, *J. Indian Math. Soc.* (1956), v. **20**, 47–87.
165. Selberg A., Discontinuous groups and harmonic analysis, Proc. ICM. Stocholm, 1962, Uppsala, Almqvist, Wikwells (1963).
166. Selberg A., On the estimation of Fourier coefficients of modular forms, *Proc. Symp. Pure Math. AMS* (1965), v. **8**, 1–15.
167. Selberg A., Recent developments in the theory of discontinuous groups of motions of symmetric spaces, Proc. 15th Scandinavian Congress, Oslo, 1968, Oslo (1969), 99–120.
168. Seminar on complex multiplication, Lecture Notes in Math., Springer (1966).
169. Serre J.–P., Lie algebras and Lie groups, New York–Amsterdam, Benjamin (1965).
170. Serre J.–P., Cours d'arithmétique, Paris, Presses Universitaires de France (1970).
171. Shimura Goro, Introduction to the arithmetic theory of automorphic functions, Princeton, Princeton University Press and Iwanami Shoten Publishers (1971).
172. Siegel C. L., Some remarks on discontinuous groups, *Ann. of Math.* (1945), v. **46**, N 4, 708–718.
173. Tate J., Fourier analysis in number fields and Hecke's zeta-function, Thesis, Princeton (1950).
174. Terras A., Harmonic analysis on symmetric spaces and applications 1, New York, Springer (1985).
175. Titchmarsh E. C., The theory of the Riemann zeta-function, Oxford (1951).
176. Vaughan R. C., Sommes trigonometriques sur les nombres premiers, *C.R. Acad. Sci., Paris*, A285 (1977), 981–983.
177. Venkov A. B., The spectral theory of automorphic functions, Proc. Intern. Congr. Math. Warszawa (1984), 909–920.
178. Vignéras M. F., Examples de sous-groups discrets non conjugués de PSL(2, **R**) qui ont même function zêta de Selberg, *C.R. Acad. Sci., Paris*, sA (1978), v. **287**: 2, 47–50.
179. Vignéras M. F., L'équation functionnelle de la fonction zêta de Selberg du groupe modulaire PSL(2, **Z**), *Soc. Math. France Aster.* (1979), v. **61**, 235–249.
180. Wallach N., On the Selberg trace formula in the case of compact quotient, *Bull. Amer. Math. Soc.* (1976), v. **82**, 171–195.
181. Warner G., Selberg's trace formula for non-uniform lattices: The R-rank 1 case, Adv. in *Math. Studies* (1979), v. **6**, 1–142.
182. Weil A., On some exponential sums, *Proc. Nat. Acad. Sci. USA* (1948), v. **34**, 204–207.
183. Weil A., Number of solutions of equations in finite fields, *Bull. AMS* (1949), v. **55**, 497–508.

184. Wolpert S., The length spectra as moduli for compact Riemann surfaces, *Ann. of Math.* (1979), v. **109**, 323–351.
185. Yang P. and Yan S.-T., Eigenvalues of the Laplacian of compact Riemann surfaces and minimal submanifolds, *Ann. Sci. Norm. Super. Pisa* (1980), v. **7**, 55–63.
186. Zograf P., Selberg trace formula for the Hilbert modular group of a real quadratic number field, *J. of Soviet Math.* (1982), v. **19**, 1637–1652.

APPENDIX 1

Monodromy Groups and Automorphic Functions

We begin with a short historic introduction, in which we describe some developments in the field of mathematics focused on the so-called Riemann–Hilbert problem. First we introduce some definitions.

The notion of monodromy is, broadly speaking, a characteristic of branching multi-valued functions. In classical mathematics, such functions appear when we solve ordinary differential equations with rational (meromorphic) coefficients or linear homogeneous systems.

Suppose we are given a system of linear equations

$$\frac{\mathrm{d}y_i}{\mathrm{d}z} \equiv y_i' = \sum_{k=1}^{n} A_{ik}(z) y_k, \quad i = 1, \ldots, n, \tag{1}$$

with rational coefficients $A_{ik}(z)$. Then, generally speaking, its solutions are not single-valued and meromorphic functions in the extended complex plane $\overline{\mathbb{C}}$. Each solution has only a finite number of isolated singularities. They lie at the poles $z_1, \ldots, z_m$ of functions $A_{ik}(z)$ and, possibly, at the point $z_{m+1} = \infty$.

We fix a fundamental system of solutions $Y(z)$ for the Equations (1). This matrix is a multi-valued function. Extending it analytically along a closed contour γ we obtain the following formula:

$$Y(z) \to Y(z) M_\gamma \tag{2}$$

with a matrix M_γ not depending on z which is called a monodromy matrix. In fact, the monodromy matrix M_γ depends only on a homotopic class $\{\gamma\}$ containing the contour γ and on the choice of the system of fundamental solutions. In addition, the mapping

$$\gamma \to M_\gamma$$

is a homomorphism of the fundamental group $\pi_1(\overline{\mathbb{C}} - \{z_1, \ldots, z_{m+1}\}) \to \mathrm{GL}(n, \mathbb{C})$, and, thus, it determines a representation of it called a monodromic one.

The branching characteristic of $Y(z)$ (2) or, more exactly, the monodromic representation $\gamma \to M_\gamma$ turns out to determine very effectively the initial system of Equations (1). As far back as 1857 Riemann supposed that for any homomorphism $\pi_1(\mathbb{C} - \{z_1, \ldots, z_m\}) \to \mathrm{GL}(n, \mathbb{C})$, where the points $z_1, \ldots, z_m$ are given in advance, there exists a system of linear homogeneous ordinary differential equations with rational coefficients of the type (1), having a fundamental system of solutions which generates the given homomorphism. In addition, one can always choose a system of Fuchsian type, i.e., a system with simplest singular points, in the role of such a system of differential equations. For the reader's convenience, we recall the definition of a singularity of Fuchsian type for the system (1). Let, for accuracy, this be the point zero. One can describe the fundamental system $Y(z)$ of Equations (1) locally (i.e., for $|z| < \epsilon$) as a product of a multi-valued function of a special form $\Phi = \exp(B \ln)$ by a one-valued matrix function Ψ which can be expanded in a Laurent series at the point zero. The point zero is called a singularity of Fuchsian type if the mentioned Laurent expansion contains only a finite number of terms with negative exponents.

This problem of Riemann's, or the Riemann–Hilbert problem, as it was referred to since Hilbert included it in his famous 'Mathematical Problems' (problem number 21) and considered its important special applications, was examined by many outstanding mathematicians including Poincaré and Schlesinger. The general, positive solution of the problem, precise from the contemporary point of view, was obtained by Plemelj who simplified and developed the Hilbert method of integral equations. In our time, this problem of Riemann's was generalized by Röhrl on the systems of differential equations, coefficients of which are meromorphic functions on an arbitrary compact or non-compact Riemann surface, and was solved by him. His proof is based on studying holomorphic vector fiber bundles over Riemann surfaces and, in fact, is the existence theorem (like the proof of Hilbert–Plemelj).

Now, putting aside the general case (see the comments to this appendix for additional information), we consider the system (1) corresponding to the equation of second order

$$\frac{d^2y}{dJ^2} + P(J)\frac{dy}{dJ} + Q(J)y = 0 \tag{3}$$

with meromorphic coefficients. The Equation (3) is classical and plays an important role in many fields of mathematics. Our interest in it is stipulated by the fact that among the monodromy groups for the Equation (3) there are Fuchsian groups of the first kind. We now refer in more detail to some classical facts on the behaviour of solutions for the Equation (3) in a neighbourhood of singular points.

Let $w = y_1 y_2^{-1}$ be a ratio of any two solutions of the Equation (3), let $\tilde{w} = \tilde{y}_1 \tilde{y}_2^{-1}$ be a ratio of another two solutions of the same Equation (3); then, evidently, w and $\tilde{w}$ are connected by a linear-fractional transformation with complex coefficients

$$\tilde{w} = \frac{aw + b}{cw + d}. \tag{4}$$

Next, suppose that the coefficients $P(J)$, $Q(J)$ of the Equation (3) are analytic in a certain region $\Omega \subset \mathbb{C}$. If, besides, Ω is simply connected then all the solutions for (3) are analytic in Ω. In the case where Ω is multiply connected, the considered equation may have solutions multi-valued in Ω, i.e. continuing analytically the solution along a closed contour going around a point which does not belong to Ω, we may return to the initial point with a new value of the solution. Formally, this new value can be easily determined. It is of great importance for us to watch a change of the ratio w of two solutions that, as is clear, can be described by a linear-fractional transformation (4). This transformation has, in addition, a non-zero determinant, if the initial solutions are linear independent which we shall assume in the sequel. The set of linear-fractional transformations of the ratio w that are obtained in this way as a result of continuing along all the closed curves forms a group which is called a group of the equation, or a monodromy group.

It is not difficult to verify that the definition mentioned above is correct, i.e., it does not depend on the choice of contours, and, up to conjugation it depends on the choice of an initial pair of solutions y_1, y_2 (of fundamental system). The monodromy group depending on the choice of coefficients P, Q may consist only of one identity transformation, may be non-trivial finite or an infinite group. In the last case, it may or may not be a discrete group in $SL(2, \mathbb{C})$.

Consider now a function $J(w)$ that is inverse to

$$w(J) = y_1(J)y_2(J)^{-1} \tag{5}$$

for some fixed fundamental system of solutions for the Equation (3). Thanks to properties of $w(J)$ mentioned above, the function $J(w)$ is automorphic with respect to the monodromy group Γ of the Equation (3), although it may not be single-valued. The single-valued condition can be stated in geometric terms and is guaranteed by the following simple theorem, the proof of which makes use only of analyticity of coefficients of the Equation (3).

THEOREM 1. (1) *If coefficients of the Equation* (3) *are analytic in a neighbourhood of the given point* J_0 *then the function* $w(J)$ *defined by formula* (5) *maps this neighbourhood of* J_0 *onto a certain domain in* $\mathbb{C}$ *in a one-to-one way.*

((2) *Let the coefficients of the Equation* (3) *be analytic in a neighbourhood of the point* J_0 *except for the point* J_0 *itself. Let* γ *be a transformation* (4) *to which the function* $w(J)$ *of the kind* (5) *is subjected in order to obtain its analytic continuing along a closed contour surrounding the point* J_0 *and not containing another singular point of the equation. Then, in order that the inverse function* $J(w)$ *be single-valued in a neigbourhood of* $w(J_0)$, *it is necessary that* γ *be either an elliptic transformation with the angle of rotation* $\frac{2\pi}{k}$, *where* k *is an integer, or a parabolic transformation.*

A class of Fuchs equations is naturally singled out in the set of all Equations (3) with meromorphic coefficients. All singular points of such equations are regular. A point a is called a regular singular point for the Equation (3) if $P(J)$, $Q(J)$ are not analytic at a, but $(J-a)P(J)$, $(J-a)^2Q(J)$ are analytic at a. Note that in a neighbourhood of a regular singular point one can easily obtain the expansions in series for solutions from the fundamental system (3). Without going into too much detail we can say that with a certain degree of accuracy the behaviour of the solutions of the equation in a neighbourhood of a regular singular point is characterized by so-called indices or solutions α_1, α_2 of a defining equation

$$\alpha^2 + (p_a - 1)\alpha + q_a = 0, \tag{6}$$

where

$$p_a = \lim_{w\to a}(w-a)P(w), \quad q_a = \lim_{w\to a}(w-a)^2 Q(w).$$

For example, one can describe the conformal properties of the function $w(J)$ (5) in terms of indices in the following way.

THEOREM 2. *In order that the function $w(J)$ maps in a one-to-one way a neighbourhood of a regular singular point a onto a certain domain in $\overline{\mathbb{C}}$, it is necessary and sufficient to fulfil one of the following conditions:*

(1) $\alpha_1 - \alpha_2 = \frac{1}{n}$ $(n \in \mathbb{Z},\ n \geq 2)$;
(2) $\alpha_1 = \alpha_2$;
(3) $\alpha_1 - \alpha_2 = 1$ *and* $r_a = \frac{1}{2}p_a s_a$, *where*

$$r_a = \lim_{J\to a}\frac{\mathrm{d}}{\mathrm{d}J}(Q(J)(J-a)^2), \quad s_a = \lim_{J\to a}\frac{\mathrm{d}}{\mathrm{d}J}(P(J)(J-a)).$$

If, in addition, J goes around a once, then $w(J)$ is subject in the case (1) to an elliptic transformation, in case (2) to a parabolic transformation and in case (3) to the identity transformation.

For many applications, it suffices to study even more specific class of the Fuchs Equations (3) with rational coefficients. Suppose we are given such an equation. We denote by $a_1, \ldots, a_n$ its singular points lying in $\mathbb{C}$ and by $\alpha_i^{(k)}$ $(i = 1,2)$ the indices at the point a_k. The regularity condition of the singular points puts a strong restriction on the functions $P(J)$, $Q(J)$. The function $P(J)$, in particular, can be expressed uniquely by means of the indices and singular points

$$P(J) = \sum_{k=1}^{n} \frac{1-\alpha_1^{(k)} - \alpha_2^{(k)}}{J - a_k}.$$

The coefficient $Q(J)$ in (3) can also be expressed through a_k and $\alpha_i^{(k)}$ but up to $n-2$ not determined parameters that one calls accessory parameters or coefficients.

We assume now that the regular singular points $a_1, \ldots, a_n$ are given and that the indices here and at infinity have the values such that in a neighbourhood of each singularity the function (5) is single-valued. The classic question arises: is it possible to choose the accessory parameters of the problem in such a way that the function $J(w)$ inverse to $w(J)$ of the form (5) is single-valued everywhere and is an elementary or a Fuchsian function? (We mention that in [1] by 'elementary' we mean rational and elliptic functions, and by Fuchsian function we mean meromorphic function on the upper half-plane H automorphic relative to any Fuchsian group.) In [1] the following existence theorem is stated.

THEOREM 3. *Under the hypothesis of Theorem 2 there exists only one differential equation such that the function inverse to the ratio of its two solutions is an elementary or a Fuchsian function.*

The proof of this theorem is based on some non-constructive ideas of the theory created for solving the uniformization problem and we do not present it here.

The problem of constructing the explicit formulas for accessory parameters in terms of groups of differential equations is still open. This problem is also important for the uniformization problem mentioned above, more precisely, in that classical proof of it attributed to Klein and Poincaré which makes use of differential equations and is not very exact from a current point of view.

This problem is commented on later, but for now we consider the Schwarz classical problem. This is formally a separate problem but, in fact, is closely connected with the aforementioned, on a differential equation for a certain conformal mapping. In this problem, using the ideas of the spectral theory of automorphic functions, one succeeds in deriving the explicit, although transcendental, formulas for the accessory coefficients.

We recall the classic statement of the problem, (see [2], [3], [4]). Let M be a simply connected polygon bounded by a finite number of arcs of the circles (a circular polygon), let H be an upper half-plane. According to the Riemann theorem, there exists a conformal mapping of open sets: $z : H \to M$; it is necessary to determine its properties. First of all the desired mapping $z(J)$ satisfies the following Schwarz equation:

$$\{z, J\} = Q_M(J), \tag{7}$$

where $\{z, J\} = z'''z'^{-1} - 3/2 \cdot z''^2 z'^{-2}$ is the Schwarz derivative, $Q_M(J)$ is the following rational function:

$$Q_M(J) = \tfrac{1}{2} \sum_{k=1}^{n} \Big(\frac{1 - \beta_k^2}{(J - a_k)^2} + \frac{c_k}{J - a_k} \Big). \tag{8}$$

In formula (8) n is the number of vertices of the polygon M, $\pi\beta_k$ is the interior angle in the vertex b_k; $a_k \in \mathbb{R} \cup \{\infty\}$, $z(a_k) = b_k$, $k = 1, \dots, n$; $c_k \in \mathbb{R}$ (c_k do not depend on J).

It is widely known that Equation (7) is closely connected with the special equation of Fuchs

$$f''(J) + \tfrac{1}{2} Q_M(J) f(J) = 0. \tag{9}$$

More precisely, if $z(J)$ is the conformal mapping from (7), then

$$z(J) = f_1(J) f_2(J)^{-1}, \tag{10}$$

where $f_1(J)$, $f_2(J)$ are linearly independent solutions to Equation (9). We now note that in the general Equation (3) considered earlier, one can exclude the term with the first derivative by a simple change of unknown function

$$y = f\exp\left(-\tfrac{1}{2}\int_{J_0}^{J} P(\tilde{J})\,\mathrm{d}\tilde{J}\right). \tag{11}$$

Equation (3) turns into Equation (9), where the expression $Q(J)-\frac{1}{2}P'(J)-\frac{1}{4}P(J)^2$ will stand instead of $\frac{1}{2}Q_M(J)$. The ratios of the solutions of (5) and (10) will coincide after the change.

We return again to formula (8). The parameters β_k, a_k, c_k are not independent from one another. The formal expansion of the Schwarz derivative $\{z, J\}$ in the neighbourhood of $J = \infty$ implies the following three simple relations:

$$\begin{gathered}\sum_{k=1}^{n} c_k = 0, \quad \sum_{k=1}^{n}(2a_k c_k + 1 - \beta_k^2) = 0, \\ \sum_{k=1}^{n}(c_k a_k^2 + a_k(1-\beta_k^2)) = 0.\end{gathered} \tag{12}$$

Counting the parameters shows that there are no other relations. In fact, the polygon M is given by $3n$ real parameters, since each of its sides is determined by the coordinates of the centre of the corresponding circle and its radius. Equation (7) determines M up to 6 real parameters

$$z \to \frac{az+b}{cz+d}, \quad \begin{pmatrix} a & b \\ c & d \end{pmatrix} \in \mathrm{SL}(2,\mathbb{C}),$$

since the Schwarz derivative does not depend on the action of a linear-fractional transformation to the function $z(J)$. $Q_M(J)$, therefore, depends on $3n-6$ real parameters.

On the other hand, Q_M contains the parameters $\beta_k, a_k, c_k \in \mathbb{R}$, $k = 1, \ldots, n$ with condition (12) which gives $3n-3$ parameters. In addition, the transformations $g : H \to H$, $g \in \mathrm{PSL}(2,\mathbb{R})$ should be taken into account since they reduce 3 parameters as yet. Thus, Q_M contains exactly $3n-6$ independent real parameters that can be, in principle, uniquely determined by the geometric parameters of the polygon. Equation (7) together with (8), (12) has as a solution the desired conformal mapping $z(J)$ for the most general circular polygon.

The real difficulty in effectively describing $z(J)$ lies not so much in the solving of Equation (7) or the simpler Equation (9), but in establishing the explicit connection between the geometric parameters of M and the coefficients of the function Q_M. If we consider the angles $\pi\beta_k$ to be determined *a fortiori* and the singular points a_k, $z(a_k) = b_k$, to be determined to some extent (up to the transformations from $\mathrm{PSL}(2,\mathbb{R})$ mentioned above), then $n-3$ parameters c_k or the accessory parameters of the problem remain unknown.

For $n = 3$ a special situation arises (we do not consider the more simple case of $n = 2$, see [1]): the equation does not contain accessory parameters and can be fully determined. We introduce some new notation. Let M_3 be a circular triangle with the angles $\pi\alpha$, $\pi\beta$, $\pi\gamma$, let the points J corresponding to the vertices be $0, 1, \infty$. The function Q_{M_3} is equal to

$$Q_{M_3}(J) = \frac{1}{2}\left(\frac{1-\alpha^2}{J^2} + \frac{1-\beta^2}{(J-1)^2} + \frac{\alpha^2+\beta^2-\gamma^2-1}{J(J-1)}\right). \tag{13}$$

Passing from Equation (9) with the function (13) to the equation with the first derivative (3) by means of the change (11) mentioned earlier, we obtain the general hypergeometric equation

$$J(1-J)f'' + (c-(a+b+1)J)f' - acf = 0 \tag{14}$$

where $a = \frac{1}{2}(1+\gamma-\alpha-\beta)$, $b = \frac{1}{2}(1-\alpha-\beta-\gamma)$, $c = 1-\alpha$.

One of the solutions to Equation (14) can be given by a hypergeometric Gauss series

$$F(a,b,c;J) = 1 + \frac{ab}{c}J + \frac{a(a+1)b(b+1)}{c(c+1)2!}J^2 + \\ + \frac{a(a+1)(a+2)b(b+1)(b+2)}{c(c+1)(c+2)3!}J^3 + \cdots, \quad |J| < 1,$$

for which an integral presentation is also known

$$F(a,b,c;J) = \frac{\Gamma(c)}{\Gamma(b)\Gamma(c-b)}\int_0^1 t^{b-1}(1-t)^{c-b-1}(1-tJ)^{-a}\,dt, \quad b>0, \quad c>b,$$

where Γ is the Euler function. For the desired transforming function $z(J)$ there exists the explicit expression; we indicate it for the triangles with $\alpha+\beta+\gamma<1$.

THEOREM 4. *The function*

$$z(J) = \frac{\int_0^1 t^{-1/2(1+\alpha+\beta+\gamma)}(1-t)^{-1/2(1+\alpha-\beta-\gamma)}(1-tJ)^{-1/2(1-\alpha-\beta+\gamma)}\,dt}{\int_0^1 t^{-1/2(1+\alpha+\beta+\gamma)}(1-t)^{-1/2(1-\alpha+\beta-\gamma)}(1-tJ)^{-1/2(1-\alpha-\beta-\gamma)}\,dt}$$

maps conformally the upper half-plane $\operatorname{Im} J > 0$ *on the circular triangle with the angles* $\pi\alpha$, $\pi\beta$, $\pi\gamma$ *provided the sum of angles is less than* π.

In order that the function $J(z)$ inverse to $z(J)$ has a unique continuation, we must, in accordance with Theorems 1, 2, lay certain conditions on the coefficients of Equation (14). Practically, these conditions mean that the corresponding monodromy group must be discrete. The necessary condition of discreteness of a group can be simply expressed, in turn, in geometric terms. The numbers α, β, γ must be reciprocal to natural ones. (Recall that $\pi\alpha$, $\pi\beta$, $\pi\gamma$ are the values of the angles of the triangle).

From the point of view of the structure of the monodromy group the case $\alpha+\beta+\gamma<1$ considered in Theorem 4 is the most non-trivial and corresponds to a Fuchsian group, in particular. In general, the monodromy group for Equation (14) can be constructed in the following way. We must take the group generated by the reflections relative to all sides of the triangle and preserve in it only products consisting of an even number of reflections. The group Γ_{M_3} thus obtained is analogous to the groups of Γ_M type we considered earlier in Chapter 8 and is called the triangle group. Corresponding automorphic functions are called triangle functions or Riemann–Schwarz functions. Certain reasoning shows that the case $\alpha+\beta+\gamma>1$ ($\alpha+\beta+\gamma=1$, $\alpha+\beta+\gamma<1$) corresponds to finite groups (to Euclidean lattices and Fuchsian groups of the first kind respectively). It can be noted here that examining the monodromy group provides an answer to a classical question dating back at least to Kummer on algebraic solutions of a hypergeometric equation. Equation (14) has algebraic solutions only for parameters $\alpha+\beta+\gamma>1$.

We now return to describing the conformal mapping $z(J)$ for a circular polygon with a number of vertices greater than 3. In order that the description by means of a differential equation be effective, it is necessary, as we have mentioned above, to clarify the nature of accessory coefficients of the problem. In the first place, it is interesting for polygons corresponding to Fuchsian monodromy groups. We now describe a method of deriving certain formulas for accessory coefficients that gives simultaneously a description of the mapping $J(z)$ inverse to $z(J)$. The method is based on ideas of the spectral theory of automorphic functions, more precisely, the mapping $J(z)$ can be obtained as the limit value of the Siegel–Selberg series which is closely connected with the resolvent of the automorphic Laplacian.

Suppose that M is a regular polygon (see Chapter 8). We shall also assume that $h=h(M)\neq 0$. For technical convenience we suppose that: (1) the number of angles of M is equal to $l=n+1$, $n\geq 2$; (2) $M\subset H=\{z\in\mathbb{C}\mid z=x+iy,\ y=\operatorname{Im} z>0\}$; (3) $b_j=z(a_j)$, $j=1,\dots,n+1$, $b_{n+1}=a_{n+1}=\infty$; (4) for any sufficiently large $a>0$ we have

$$\{z\in M\mid \operatorname{Im} z\geq a\}=[0,\tfrac{1}{2}]\times[a,\infty).$$

The inverse mapping (the Klein absolute invariant) $J(z)=J_M(z)$ exists, it satisfies the differential equation

$$-\{J_M(z),z\}=J_M'^2(z)Q_M(J_M(z)) \tag{15}$$

and is a Γ_M-automorphic function analytic in H. Furthermore, Γ_M is a group defined in Chapter 8 by a regular polygon M. The group contains a cyclic subgroup Γ_∞ generated by a shift $z\to z+1$. We have a Fourier extension

$$J_M(z)=\sum_{\substack{k\in\mathbb{Z}\\ k\geq -1}} A_k\exp(2\pi ikz). \tag{16}$$

The following formula is widely known and can be derived easily:

$$\begin{aligned} Q_M(J) &= \prod_{k=1}^{n}(J-a_k)^{-1}\Big(E_{n-2}(J)+\sum_{k=1}^{n} N_k(J-a_k)^{-1}\Big), \\ N_k &= \tfrac{1}{2}(1-\beta_k^2)\prod_{l=1}^{n}{}' (a_k-a_l). \end{aligned} \tag{17}$$

Here the prime means that the product is taken over all l except for $l=k$; then $E_{n-2}(J)=\sum_{k=0}^{n-2} X_k J^k$, $X_{n-2}=\frac{1}{2}$. The problem is in finding the coefficients X_k, $k=0,\dots,n-3$.

We introduce a notation: $\prod_{k=1}^{n}(J-a_k)=\sum_{k=0}^{n} B_k J^k$, $B_n=1$, $J=J(z)$. Next

$$\begin{aligned} \Phi_p(r) &= \sum_{\substack{t_1+\cdots+t_p+s_1+\cdots+s_4=r \\ -1\le t_1,\dots,t_p,s_1,\dots s_4<\infty}} A_{t_1}\cdots A_{t_p}A_{s_1}\cdots A_{s_4}s_1\cdots s_4, \\ \Psi(r) &= \sum_{k=0}^{n} B_k \sum_{\substack{l_1+\cdots+l_k+m_1+m_2=r \\ -1\le l_1,\dots,l_k,m_1,m_2<\infty}} A_{l_1}\cdots A_{l_k}A_{m_1}A_{m_2}m_1^2 m_2\big(\tfrac{3}{2}m_2-m_1\big). \end{aligned} \tag{18}$$

In (18) we suppose that $-4\ge r\ge -n-2$, $0\le p\le n-2$.

THEOREM 5. (1) *The matrix* $\{\Phi_p(r)\}$ *is invertible. What is more, it is triangular with positive elements on the diagonal.*

(2) *For the coefficients* X_k, $0\le k\le n-3$ *the following formula is valid*

$$X_k=\sum_{r=-n-2}^{-4}\eta_k(r)\Psi(r), \tag{19}$$

where $\{\eta_p(r)\}=\{\Phi_p(r)\}^{-1}$.

The proof of this theorem is based on formulas (15)–(17).

Formula (19) makes sense if the coefficients A_k from (15) are known. We now obtain certain formulas for $A_{-1}^{-1}A_k$, $k\ge 1$. Here the circular Hardy–Ramanujan–Littlewood–Rademacher–Lehner method could be useful (see [5]) but currently it is not sufficiently accurate for an arbitrary group Γ_M. So we shall use another method based on the connection between the resolvent of the operator $A(\Gamma_M;1)$ and the Green harmonic function. In addition, this method gives more information on $J_M(z)$.

Let $G_M(z,z')$ be the Green function for the Dirichlet boundary problem on M for the ordinary Laplace operator. From conformal invariance of the Green function and Riemann theorem we have

LEMMA 1. *The following formula is valid:*

$$G_M(z,z')=-\tfrac{1}{2\pi}\ln\left|\frac{J_M(z)-J_M(z')}{J_M(z)-\overline{J_M(z')}}\right|,$$

where the bar means complex conjugation.

Let $r(z, z'; s; \Gamma_M)$ be the kernel of the resolvent $(A(\Gamma_M; 1) - s(1-s))^{-1}$ at some regular point $s(1-s) \in \mathbb{C}$ (for example, $\operatorname{Re} s > 1$; see Chapter 4).

THEOREM 6. *The following formula is valid:*

$$G_M(z, z') = \lim_{s \to 1+0} (r(z, z'; s; \Gamma_M) - r(-\bar{z}, z'; s; \Gamma_M)).$$

Now we introduce the Siegel–Selberg series:

$$F_n(z, s; \Gamma_M) = \sum_{\gamma \in \Gamma_\infty \backslash \Gamma_M} \exp(2\pi i n x(\gamma z)) \sqrt{y(\gamma z)} I_{s-1/2}(2\pi |n| y(\gamma z))$$

that is absolutely convergent for $\operatorname{Re} s > 1$. Here $n \in \mathbb{Z}$, $z = \operatorname{Re} z + i \operatorname{Im} z = x(z) + iy(z)$, I_s is the Bessel (or modified) function of imaginary argument. In what follows, the ordinary Bessel function $J_s(z)$ also appears. Unfortunately, a standard notation for it coincides with standard notation $J_M(z)$ for the Klein invariant. So for contrast, we shall omit the index M in $J_M(z)$ supposing M to be fixed.

Theorem 5.4 implies the following lemma.

LEMMA 2. *Let $m, n \in \mathbb{Z}$ $(m, n \neq 0)$, then the m-th coefficient of the Fourier series $F_n(z, s)$ is equal to*

$$\begin{aligned} &\delta_{mn} \sqrt{y} I_{s-1/2}(2\pi |n| y) + 2\sqrt{y} K_{s-1/2}(2\pi |m| y) \times \\ &\times \sum_{\substack{\gamma \in \Gamma_\infty \backslash \Gamma_M / \Gamma_\infty \\ c > 0}} \tfrac{1}{c} S(-m, -n; c) M_{2s-1}\left(\tfrac{4\pi}{c} \sqrt{-mn}\right), \end{aligned} \tag{20}$$

where

$$M_{2s-1}(a\sqrt{-mn}) = \begin{cases} I_{2s-1}(a\sqrt{|mn|}), & mn < 0, \\ J_{2s-1}(a\sqrt{|mn|}), & mn > 0. \end{cases}$$

In these formulas we accept the following notation: $K_s(z)$ is a modified Bessel function, $S(m, n; c)$ is a general Kloosterman sum

$$S(m, n; c) = \sum_{0 \leq d < c} \exp\left(2\pi i \frac{ma + nd}{c}\right),$$

where $\gamma z = (az + b)(cz + d)^{-1})$, γ is taken in the class $\Gamma_\infty \backslash \Gamma_M / \Gamma\infty$. (The sum is determined uniquely by the matrix element c and the parameters m, n.)

We now proceed to deriving the formulas for the coefficients $A_k A_{-1}^{-1}$.

THEOREM 7. *The following formula holds:*

$$\operatorname{Im}(J(z) A_{-1}^{-1}) = i\pi \lim_{s \to 1+0} (F_1(z, s) - F_{-1}(z, s)).$$

We outline the proof of this theorem. To prove this, it is sufficient to find the main term of the asymptotic expansion of the Green function $G_M(z, z')$ for large values of $\operatorname{Im} z$ in two various ways according to Lemma 1 and Theorem 6 and to compare the results thus obtained.

On the one hand, we have

$$-\frac{1}{2\pi}\ln\left|\frac{J(z)-J(z')}{J(z)-\overline{J(z')}}\right| = \frac{\operatorname{Im} J(z')}{\pi A_{-1}e^{2\pi y}}\sin 2\pi x + O(e^{-2\pi y}), \tag{21}$$

$$y \to \infty, \quad y = \operatorname{Im} z.$$

We now recall certain results from the spectral theory of automorphic functions concerning the Fourier expansion for the kernel of the resolvent $r(z,z';s;\Gamma_M)$ (see Chapter 5). In the half-plane $\operatorname{Re} s > 1$ the kernel of the resolvent admits the presentation in the form of an absolutely convergent series by a discrete group

$$r(z,z';s;\Gamma_M) = \sum_{\gamma\in\Gamma_M} k(z,\gamma z';s),$$

where $k(z,z';s)$ is the Green function of the operator $L - s(1-s)$ on the entire Lobachevsky plane H. Similar expansion may be written, making use of the Green function $w(z,z';s)$ of the same operator, but on the fundamental domain of the group Γ_∞ in H. Thus:

$$r(z,z';s;\Gamma_M) = \sum_{\gamma\in\Gamma_\infty\backslash\Gamma_M} w(z,\gamma z';s). \tag{22}$$

Separating the variables in the equation $Lu - \lambda u \equiv -y^2(u_{xx}+u_{yy})^{-\lambda u} = f$, we obtain

$$w(z,z';s) = t(y,y';s) + 2\sum_{k=1}^{\infty}\cos(2\pi k(x-x'))m_k(y,y';s), \tag{23}$$

where $z = x+iy$, $z' = x'+iy'$,

$$t(y,y';s) = \frac{1}{2s-1}\begin{cases} y^s y'^{1-s}, & y' \geq y,\\ y^{1-s}y'^s, & y \geq y',\end{cases}$$

$$m_k(y,y';s) = \begin{cases} \Psi_1^{(k)}(y')\Psi_2^{(k)}(y), & y' \geq y,\\ \Psi_1^{(k)}(y)\Psi_2^{(k)}(y'), & y \geq y',\end{cases}$$

and $\Psi_1(y) = \sqrt{y}K_{s-1/2}(2\pi|k|y)$, $\Psi_2(y) = \sqrt{y}I_{s-1/2}(2\pi|k|y)$. Let $z,z' \in F_M$, $y = \operatorname{Im} z \to \infty$. Then $y > y'$, $y > y(\gamma z')$ for any $\gamma \in \Gamma_\infty\backslash\Gamma_M$. So the main term in the asymptotic expansion of the difference $r(z,z';s;\Gamma_M) - r(-\bar{z},z';s;\Gamma_M)$ as $y\to\infty$ is generated by only one summand of the right-hand side of (23) corresponding to $k=1$, and this term is equal to

$$2\sum_{\gamma\in\Gamma_\infty\backslash\Gamma_M}(\cos(2\pi(x-x(\gamma z'))) - \cos(2\pi(x+x(\gamma(z')))))m_1(y,y(\gamma z');s)$$

$$= \sqrt{y}K_{s-1/2}(2\pi y)\,4\sin(2\pi x)\,\frac{1}{2i}(F_1(z',s)-F_{-1}(z',s)) \underset{y\to\infty}{\sim} \tag{24}$$

$$\sim e^{-2\pi y}\sin(2\pi x)\,\tfrac{1}{i}(F_1(z',s;\Gamma_M)-F_{-1}(z',s;\Gamma_M));$$

we use the known asymptotic expansion for the function $K_s(z)$. The assertion of the theorem now follows from Lemma 1, Theorem 6, formulas (21), (24).

THEOREM 8. *The following formulas for the Fourier coefficients of the Klein invariant $J(z)$ are valid:*

$$A_kA_{-1}^{-1}=\frac{2\pi}{\sqrt{k}}\sum_{c>0}\tfrac{1}{c}\Big(S(-k,1;c)I_1\Big(\frac{4\pi\sqrt{k}}{c}\Big)-$$

$$-S(-k,-1;c)J_1\Big(\frac{4\pi\sqrt{k}}{c}\Big)\Big)\quad k\ge 2 \tag{25}$$

$$A_1A_{-1}^{-1}=2\pi\sum_{c>0}\tfrac{1}{c}\Big(S(-1,1;c)I_1\Big(\frac{4\pi}{c}\Big)-S(-1,-1;c)J_1\Big(\frac{4\pi}{c}\Big)\Big)+1. \tag{26}$$

The assertion of the theorem easily follows from Lemma 2 and Theorem 7. Here one must only use the fact that the coefficients A_k are real and also use the following well known formulas

$$\sqrt{y}K_{1/2}(2\pi y)=\tfrac{1}{2}e^{-2\pi y},\quad \sqrt{y}I_{1/2}(2\pi y)=\tfrac{1}{\pi}\mathrm{sh}(2\pi y)$$

REMARK. The second summands in the sums in (25), (26) are closely connected with the Fourier coefficients of the first Poincaré series of weight 2 for the group Γ_M. In fact, the general mth Poincaré series of weight $l\ge 2$ is defined by the formula

$$P_m(z,l;\Gamma_M)=m^{l-1}i^{-l}\sum_{\gamma\in\Gamma_\infty\backslash\Gamma_M}\frac{e^{2\pi i m\gamma z}}{(cz+d)^l},$$

where the series converges absolutely for $l>2$; its convergence for $l=2$ can also be proved. The Fourier coefficient with the number $n>0$ of the Poincaré series P_m is equal to

$$c_{mn}=(mn)^{(l-1)/2}\Big(i^{-l}\delta_{mn}+2\pi\sum_{\substack{\gamma\in\Gamma_\infty\backslash\Gamma_M/\Gamma_\infty\\ c>0}}c^{-1}S(m,n;c)J_{l-1}\Big(\frac{4\pi\sqrt{mn}}{c}\Big)\Big), \tag{27}$$

where δ_{mn} is the Kronecker symbol. The connection between Equations (25)–(27) becomes evident if we set $m=1$, $l=2$ in (27), make use of the known property of the general Kloosterman sum $S(m,n;c)=S(-n,-m;c)$ and, finally, if we notice that the summation is taken in all three of these formulas over the same set of numbers c.

We note that each of the groups Γ_M has genus zero, so there do not exist non-trivial cusp forms of weight two for them, including the Poincaré series with $m\neq 0$.

Hence, it follows that for any group generated by reflections there are no summands with $S(-k,-1;c)$, $S(-1,-1;c)$ in sums (25), (26). Thus, we have the assertion.

THEOREM 9. *For Fourier coefficients of the Klein invariant the following formula is valid:*

$$A_k A_{-1}^{-1} = \frac{2\pi}{\sqrt{k}} \sum c^{-1} S(-k,1;c) I_1\left(\frac{4\pi\sqrt{k}}{c}\right), \quad k \geq 1.$$

It can be noted that the formula of Theorem 9 generalizes the Rademacher formula obtained by him for the modular group (see [6]).

The Fourier coefficients A_0, A_{-1} of the invariant J_M which are beyond our consideration cannot be found in such a way. In some senses these coefficients are determined by the conditions of the normalization of $J_M(z)$. The fact is that we define the invariant $J_M(z)$ as a conformal mapping, and such a definition does not determine $J_M(z)$ uniquely.

Commentary

We direct the reader who is interested in current theory of ordinary differential equations connected with the Riemann–Hilbert problem to the survey [7].

The classical monodromy theory for equations of the second order considered above is presented in [1]. Theorems 1, 2, 3 are taken from there.

The uniformization problem is as follows: to present an arbitrary Riemann surface S as a quotient $S = \Gamma\backslash H$ of a corresponding universal covering H by a discrete group Γ and to find the canonical mapping $\sigma : H \to S$. Thus, to each analytic function f on S there corresponds a one-to-one analytic function $f \circ \sigma$ on the universal covering.

The method of solving the uniformization problem attributed to Klein and Poincaré is in the correct choice of $3g-3$ accessory parameters, where g is the genus of S ($g \geq 2$), in the equation

$$\frac{\mathrm{d}^2 U}{\mathrm{d}x^2} + R(x,y)U = 0, \quad P(x,y) = 0,$$

where $R(x,y)$ is a certain rational function, and $P(x,y)$ is a polynomial. Next, having the solutions to this equation, one can construct a canonical mapping σ and thus solve the uniformization problem in a sufficiently effective way, as it seems at first sight.

Although it is now considered good form to criticize the classics, we are not going to do so here, instead directing the reader to further information on the connection of the uniformization problem with the differential equations to the survey by Hejhal [8].

We would also like to mention the monograph by Nehari [4], in which the classical results on determining the accessory parameters in the considered problem on a conformal mapping are very well described. In particular, the reader can find there certain information on the hypergeometric equation (Theorem 4).

Theorems 5–9 are attributed to the author (see [9]). The definition of the Siegel–Selberg series is in Niebur [10]. The information on the spectral theory of automorphic functions that is necessary for proving Theorem 7 can be found in [11].

References

1. Ford L. R., Automorphic functions, New York, McGraw-Hill (1929).
2. Schwarz H. A., Gesammelte mathematische Abhandlungen, V. 2, Berlin, Springer (1890).
3. Hurwitz A., Courant R., Funktionentheorie, Berlin, Springer, 1964.
4. Nehari Z., Conformal mapping, New York, McGraw-Hill (1952).
5. Lehner J., Discontinuous groups and automorphic functions, AMS, 8 Providence (1964).
6. Rademacher H., The Fourier coefficients of the modular invariant $J(z)$, *Amer. J. Math.* (1938), v. **60**, 501–512.
7. Arnoldt V. I., Iljashenko Yu. S., Ordinary differential equations, INiT, Sovrem. problemy matematiki, *Fundament. napravlenija*, (1985), v. **1**, 7–150 (Russian).
8. Hejhal D. A., Monodromy groups and linearly polymorphic functions, *Acta Math.* (1975), v. **135** , N 1–2, 1–55.
9. Venkov A. B., On exact formulas for accessory coefficients in the Schwarz equation, *Functional Analysis and its Applications* (1983), v. **17**, N 3, 1–8 (Russian).
10. Niebur D., A class of non-analytic automorphic functions, *Nagoya Math. J.* (1973), v. **52**, 133–145.
11. Venkov A. B., Spectral theory of automorphic functions, *Proceedings of Steklov Inst. Math.* (1982), N 4, 1–163.

APPENDIX 2

Automorphic Functions for Effective Solutions of Certain Issues of the Riemann–Hilbert Problem*

Introduction

A recent revival in interest in classical mathematics is the main motivation for writing this part of the Appendix. As can be seen from the title, the main issue is the problems that were the focus of attention for mathematicians eighty to a hundred years ago. One can find the roots of these problems in the remarkable works of Klein and Fricke [1], [2].

As time has passed, we now have the opportunity to expand classical theory with more details, to illustrate ideas with a large number of examples. Here, we examine the Schwarz equation connected with the hypergeometric equation and its change of variables via special algebraic functions. At the same time, this is a new survey of results in the theory of discrete groups, important for this field. Not all these facts can be utilized immediately, but they are included in this survey as we consider them to be significant for possible future development of the theory.

A precise statement of the problem which is solved in this part can be found in Section 3. Sections 1 and 2 are introductory and the information they contain is classical.

1. Fuchsian Groups

The Fuchsian group of the first kind Γ is characterized by the finiteness condition of the volume of its fundamental domain relative to an invariant measure on the upper half-plane H. This is a finitely generated discrete subgroup of the group $\mathrm{PSL}(2,\mathbb{R})$, i.e., the group of conformal homeomorphisms of H. The group Γ is given by the following generators:

$$\begin{aligned} &A_1, A_2, \ldots, A_g, B_1, B_2, \ldots, B_g \quad \text{are hyperbolic,} \\ &V_1, V_2, \ldots, V_h \quad \text{are parabolic,} \\ &E_1, E_2, \ldots, E_k \quad \text{are elliptic;} \end{aligned} \tag{1}$$

---

* This appendix is a slightly modified version of a paper originally published in 'Zapiski Nauchnykh Seminarov LOMI' **162** (1987), pp. 5–42.

and by the following relations:

$$\begin{aligned} &[A_1, B_1] \cdots [A_g, B_g] E_1 \cdots E_k V_1 \cdots V_h = 1 \\ &E_1^{n_1} = E_2^{n_2} = \cdots = E_k^{n_k} = 1. \end{aligned} \tag{2}$$

In this case Γ has the signature $(g, n_1, \ldots, n_k; h)$.

It is widely known that any elliptic (parabolic) element of the group Γ is conjugate to a power of one of $E_j(V_j)$ in Γ and none of the powers of the generators (1) are conjugate in Γ.

The fundamental domain $F = H/\Gamma$ that is compactified by adding h-points corresponding to fixed points of parabolic transformations is provided with a natural analytic structure and turns into a compact Riemann surface $\widehat{F}$ of genus g. (While introducing a complex structure on F, particular attention is given to neighbourhoods of fixed points of elliptic and parabolic generators).

The volume of the fundamental domain F is computed by the Gauss–Bonnet formula

$$|F| = 2\pi\Big[2g - 2 + h + \sum_{j=1}^{k}(1 - 1/n_j)\Big]. \tag{3}$$

In this part we shall consider Γ and F mainly with the condition $g = 0$.

2. Differential Equations

Let $J : F \to \mathbb{C}$ be a conformal mapping of the fundamental domain of the group Γ from Section 1 with the condition $g = 0$ onto a complex line that is expanded (with the same notation) as an automorphic function, Hauptfunktion, onto the entire upper half-plane H. In other words, there exists an isomorphism of complex manifolds $\widehat{F} \to \widehat{\mathbb{C}}$, where $\widehat{\mathbb{C}}$ is the Riemann sphere, which induces an univalent automorphic function $J(z)$ on H. The function $J(z)$ is defined up to a linear-fractional transformation and satisfies a Schwarz equation

$$-\{J, z\} = J'^2(z) Q_\Gamma(J(z)). \tag{4}$$

The following equality is valid for the inverse mapping $z(J)$

$$\{z, J\} = Q_\Gamma(J). \tag{5}$$

In formulas (4), (5) $\{J, z\}$, $\{z, J\}$ are the Schwarz derivatives, $\{z(J), J\} = z'''(J)z'(J)^{-1} - 3/2 \cdot z''(J)z'^{-2}(J)$, the prime means differentiation on the argument of a function, $Q_\Gamma(J)$ is a certain rational function. The fact that there appears a rational function can be determined by the properties of invariance of the Schwarz derivative

$$\{\gamma z, J\} = \{z, J\}, \quad \gamma \in \Gamma.$$

Hence it follows that the left-hand side of equality (5) is analytic outside the poles of an automorphic function and, thus, it is rational of a univalent function. Formula (4) is a simple consequence of (5).

We now proceed to describing the function $Q_\Gamma(J)$. Certain arguments show that $Q_\Gamma(J)$ is analytic everywhere except for the values of $J(z)$ in which this function is ramified. If a is the image of the branching point of order m of the function $J(z)$ then in a neighbourhood of $J = a$ the following asymptotic formula holds:

$$Q_\Gamma(J) = \tfrac{1}{2}\left(1 - \frac{1}{m^2}\right)(J-a)^{-2}(1+o(1)). \tag{6}$$

Supposing that the infinity is not the image of the branching point of the function J which can always be done by means of a linear-fractional transformation, we obtain

$$Q_\Gamma(J) = O(J^{-4}), \quad J \to \infty. \tag{7}$$

The enumerated properties enable one to compute $Q_\Gamma(J)$ up to some parameters. We introduce the following notation: b_j is an elliptic vertex of the fundamental domain F corresponding to the generator E_j $(1 \le j \le k)$ and b_j is a cusp of the domain F corresponding to the generator V_l, where $1 \le l \le h$, $k + l = j$, $k+1 \le j \le h+k = n$. In addition,

$$m_j = \begin{cases} n_j, & 1 \le j \le k, \\ \infty & k+1 \le j \le n. \end{cases} \tag{8}$$

Then

$$Q_\Gamma(J) = \tfrac{1}{2}\sum_{j=1}^{n} \frac{1 - 1/m_j^2}{(J-a_j)^2} + \sum_{j=1}^{n} \frac{c_j}{J - a_j}, \tag{9}$$

$a_j = J(b_j)$, $1 \le j \le n$. The parameters c_j do not depend on J and satisfy the conditions

$$\begin{gathered} \sum_{j=1}^{n} c_j = 0, \quad \sum_{j=1}^{n}[2a_jc_j + 1 - 1/m_j^2] = 0, \\ \sum_{j=1}^{n}[c_ja_j^2 + a_j(1 - 1/m_j^2)] = 0, \end{gathered} \tag{10}$$

which follows from (7).

Gaining further information on parameters c_j and, hence, on the function $Q_\Gamma(J)$ demands more detailed analysis, and before discussing it we recall that the Fuchs linear equation

$$f''(J) + \tfrac{1}{2}Q_\Gamma(J)f(J) = 0 \tag{11}$$

is closely connected with the Schwarz equations (4), (5). Namely, if $f_1(J)$, $f_2(J)$ are linear independent solutions of Equation (11) then the quotient $f_1(J)f_2(J)^{-1}$

is the solution of Equation (5). Equation (11) has exactly n singular points which is equal to the total number of parabolic and elliptic vertices of the fundamental domain F corresponding to the canonical choice of generators of $\Gamma(1)$. All these points are regular singular and Equation (11) belongs to the class of the Fuchs equations. In addition, the group Γ will be a monodromy group of Equation (11) for the mentioned choice of the function $Q_\Gamma(J)$.

3. The Riemann–Hilbert Problem for Fuchsian Groups of the First Kind

Hilbert's twenty-first problem or the Riemann–Hilbert problem is as follows: show that a linear differential equation of the Fuchsian type with given singular points and given monodromy group always exists. In a form of existence theorem this general problem was solved by the efforts of many outstanding mathematicians: Hilbert, Plemelj, Birkhoff and Röhrl. After proving the existence theorem, the natural question of an effective solution to this problem arises (see Lappo–Danilevskiĭ [3]), i.e. on constructing the differential equation in terms of the monodromy group. (We see that the term 'construction' is not mathematical and the effectiveness of the solution may be understood by different individuals in different ways). For general monodromy groups this question is very difficult for what takes place in the case of the Fuchsian group, from Section 1 with the natural restriction $g = 0$ (i.e. the question is of differential equations on the Riemann sphere).

Thus, the construction of the function $Q_\Gamma(J)$ in terms of the problem we are investigating here is a special case of solving the Riemann–Hilbert problem in an effective way. It is clear that for constructing $Q_\Gamma(J)$ it is sufficient to construct in terms of Γ a Hauptfunktion $J(z)$ and an inverse function $z(J)$, since by Equation (4) we have

$$Q_\Gamma(J(z)) = \tfrac{3}{2}J''^2(z)J'^{-4}(z) - J'''(z)J'^{-3}(z),$$
$$Q_\Gamma(J) = Q_\Gamma(J(z(J))).$$

These are the principal results of [4]. Introduce the following notation: we denote by

$$F(z,s;\Gamma) = \sum_{\gamma\in\Gamma_\infty\backslash\Gamma} \sqrt{y(\gamma z)}I_{s-1/2}(2\pi y(\gamma z))\exp(-2\pi i x(\gamma z)), \tag{12}$$

a special Poincaré series, the Siegel–Selberg series, where Γ is as in Section 1 with the condition $g = 0$, $h \geq 1$. We suppose that ∞ is a cusp of the fundamental domain F of the group Γ, $\Gamma_\infty \subset \Gamma$ is a subgroup holding fixed this cusp, Γ_∞ is generated by $V: z \to z+1$, $z \in H$. It can be done if we replace Γ by its conjugate in $\mathrm{PSL}(2,\mathbb{R})$. Next, in (12) $z = x+iy$, $\gamma z = x(\gamma z) + iy(\gamma z)$, $I_s(z)$ is the modified Bessel function. The series (12) converges absolutely as $\mathrm{Re}\, s > 1$ and defines a special non-analytic function automorphic relative to the group Γ.

THEOREM 1. *There exists a finite limit as $s \to 1+0$ of the function $F(z,s;\Gamma)$ equal to*

$$\lim_{s\to 1} F(z,s;\Gamma) = J(z).$$

Thus, we represented Hauptfunktion $J(z)$ as a series over the discrete group Γ. The corresponding theorem for the function $Q_\Gamma(J)$ (more precisely, for the parameters c_j (see (9)) is proved in [4], [5], [6]. It is presented in a somewhat schematic form.

THEOREM 2. *Let the function F be as in Theorem 1.*

(1) *Each parameter c_j of the function Q_Γ can be expressed by the known rational function of a finite number of variables, namely: of the Fourier coefficients A_k for the function $J(z)$ relative to the subgroup Γ_∞ and of the points $a_i = J(b_i)$.*

(2) *Each Fourier coefficient A_k can be represented as an infinite series in the matrix elements of the group Γ with coefficients that are products of general Kloosterman sum by the Bessel function.*

(3) *At each cusp b_i the value of $J(b_i)$ is an infinite series in the matrix elements of the group Γ with the Ramanujan sums as coefficients.*

Unfortunately, the series over the discrete group Γ in Theorems 1, 2 are not absolutely convergent, and, in addition, the matrix elements of general Fuchsian groups can hardly be determined which does not allow one to use them effectively. Furthermore, such a transcendence of the formulas in Theorems 1 and 2 seems to be the crux of the matter. The experience of the classical theory of automorphic functions shows that for general Fuchsian groups Γ there cannot be expected to be simple expressions for the Fourier coefficients A_k of the function $J(z)$ and for the numbers $a_i = J(b_i)$.

We further describe a class of groups Γ for which the problem of effectively determining the Fourier coefficients A_k of the functions $J(z)$ and, hence, the problem of determining the function $Q_\Gamma(J)$ itself is real, but first we have to carry out certain auxiliary investigations.

4. Equations with Three Singular Points

For $n = 3$ Equation (5) has the form

$$\begin{aligned}\{z,J\} = {} & \frac{1-1/m_1^2}{2}\frac{(a_1-a_2)(a_1-a_3)}{(J-a_1)^2(J-a_2)(J-a_3)} + \\ & + \frac{1-1/m_2^2}{2}\frac{(a_2-a_1)(a_2-a_3)}{(J-a_2)^2(J-a_1)(J-a_3)} + \\ & + \frac{1-1/m_3^2}{2}\frac{(a_3-a_1)(a_3-a_2)}{(J-a_3)^2(J-a_1)(J-a_2)},\end{aligned} \tag{13}$$

unknown (accessory) parameters c_j being absent in this case. As indicated above, the transformation $\gamma J(z)$, $\gamma \in \mathrm{SL}(2,\mathbb{C})$ is also conformal along with $J(z)$. We now suppose that $J(z)$ is fixed by the following three conditions

$$J(b_1) = a_1 = 0, \quad J(b_2) = a_2 = 1, \quad J(b_3) = a_3 = \infty$$

(standard normalization). Equation (13) turns into

$$\{z, J\} = \frac{1 - 1/m_1^2}{2J^2} + \frac{1 - 1/m_2^2}{2(1-J)^2} + \frac{1/m_1^2 + 1/m_2^2 - 1/m_3^2 - 1}{2J(J-1)}. \tag{14}$$

The right-hand side of (14) will be denoted by $Q(J; m_1, m_2, m_3)$. The corresponding Fuchs equation (11) is equal to

$$f''(J) + \tfrac{1}{2} Q(J; m_1, m_2, m_3) f(J) = 0 \tag{15}$$

which is equivalent to the hypergeometric equation

$$g''(J) + \frac{(\alpha+\beta+1)J - \gamma}{J(J-1)} g'(J) + \frac{\alpha\beta}{J(J-1)} g(J) = 0 \tag{16}$$

for the parameters

$$1/m_1 = 1 - \gamma, \quad 1/m_2 = \gamma - \alpha - \beta, \quad 1/m_3 = \beta - \alpha. \tag{17}$$

The change of variables that connects (15) and (16) has the form

$$g(J) = f(J) \exp\Bigl(-\tfrac{1}{2} \int_{J_0}^{J} P(\tilde{J})\, \mathrm{d}\tilde{J}\Bigr),$$

where J_0 is any regular point and the function $P(J)$ is the coefficient of $g'(J)$ in (16). We note that the equality holds for the quotients of partial solutions to the Equations (15), (16)

$$f_1(J) f_2^{-1}(J) = g_1(J) g_2^{-1}(J).$$

The connection with the hypergeometric equation allows one to construct the solution to the Schwarz equation (14) $z(J)$ in terms of a hypergeometric function. It is known that one of the solutions of (16) is given by the Gauss hypergeometric series

$$F(\alpha,\beta,\gamma;J) = 1 + \frac{\alpha\beta}{\gamma} J + \frac{\alpha(\alpha+1)\beta(\beta+1)}{\gamma(\gamma+1)2!} J^2 + \\ + \frac{\alpha(\alpha+1)(\alpha+2)\beta(\beta+1)(\beta+2)}{\gamma(\gamma+1)(\gamma+2)3!} J^3 + \cdots, \quad |J| < 1.$$

The recognized formulas for transforming the hypergeometric function, which we shall denote by $F(\alpha,\beta,\gamma;J)$, enable one to examine it everywhere outside the singular points $J = 0$, $J = 1$, $J = \infty$. The following integral presentation is useful:

$$F(\alpha,\beta,\gamma;J) = \frac{\Gamma(\gamma)}{\Gamma(\beta)\Gamma(\gamma-\beta)} \int_0^1 t^{\beta-1}(1-t)^{\gamma-\beta-1}(1-Jt)^{-\alpha}\, dt, \tag{18}$$

$$\beta > 0, \quad \gamma > \beta,$$

where Γ is the Euler function. The desired solution $z(J)$ is equal to

$$z(J) = \frac{F(\alpha,\beta,\gamma;J)}{F(\alpha,\beta,1+\alpha+\beta-\gamma;1-J)} \tag{19}$$

up to a normalizing factor independent of J. This function realizes a conformal mapping of the upper half-plane $\operatorname{Im} J > 0$ onto a geodesic triangle in H with the angles π/m_1, π/m_2, π/m_3 (see (17)). Notice that for both hypergeometric functions from (19) the integral presentation (18) is valid. Formula (19) is written for differential equations of generic position. In our case this means the following. We assume that for Equation (16) the difference of the roots of defining algebraic equations at singular points $J = 0$, $J = 1$, $J = \infty$ is different from zero. These differences are equal to

$$1-\gamma = 1/m_1, \quad \gamma-\alpha-\beta = 1/m_2, \quad \beta-\alpha = 1/m_3,$$

i.e., the compact triangles in H correspond to differential equations of generic position.

In the particular case when the difference of the roots of at least one defining equation is zero, the pair of linearly independent solutions to Equation (16) in the neighbourhood of the corresponding singular point looks somewhat different. For example, for the triangle with the angles π/q, $\pi/2$, 0 ($m_1 = q$, $m_2 = 2$, $m_3 = \infty$) (see [7])

$$\begin{aligned} g_1(J) &= J^{-\alpha}[F_1(\alpha,\alpha-\gamma+1;J^{-1}) - \ln J \cdot F(\alpha,\alpha-\gamma+1,1;J^{-1})], \\ g_2(J) &= J^{-\alpha}F(\alpha,\alpha-\gamma+1,1;J^{-1}), \end{aligned} \tag{20}$$

where $\alpha = \beta = \frac{1}{2}(\frac{1}{2} - 1/q)$, $\gamma = 1 - 1/q$, F_1 is a modified hypergeometric series in powers of J^{-1} the coefficients of which can easily be defined by the substitution in (16).

5. Additional Information on Fuchsian Groups

Let Γ_1, Γ_2 be two Fuchsian groups of the first kind. They are commensurable iff $\Gamma_1 \cap \Gamma_2$ has finite index in Γ_1 and Γ_2. It is clear that commensurability is an equivalence relation. We define a commensurator of Γ in $\mathrm{PSL}(2,\mathbb{R})$ by $\mathrm{Comm}(\Gamma) = \{g \in \mathrm{PSL}(2,\mathbb{R}) \mid \Gamma \text{ is commensurable with } g\Gamma g^{-1}\}$. The commensurator $\mathrm{Comm}(\Gamma)$ is the group containing Γ.

The class of groups Γ, each of which is commensurable with some triangular group, is important in what follows. Recall the definition and some properties of the triangular groups.

A triangular group is, by definition, a Fuchsian group of the first kind from Section 1 with $g = 0$ and three generators. These generators may be parabolic and elliptic in any proportion, i.e. $h + k = 3$. We denote the signature of the triangular group by a triple of numbers m_j (see (8)) (m_1, m_2, m_3).

As any Fuchsian group, the triangular group is determined by the signature uniquely up to conjugation in $\mathrm{PSL}(2,\mathbb{R})$. The parameters can always be ordered as follows:

$$2 \leq m_1 \leq m_2 \leq m_3 \leq \infty.$$

We note that the Gauss–Bonnet formula (3) gives the relation

$$1/m_1 + 1/m_2 + 1/m_3 < 1.$$

The maximality property is the significant property of the triangular group.

THEOREM 3. (see [8]). *Let Δ be a discrete group of conformal isometries of the half-plane H. If Δ has a triangular group Δ_0 as a subgroup then Δ is triangular.*

We indicate the best known examples of the triangular groups. These are the Hecke groups $G(2\cos\frac{\pi}{q})$ with signature $(2, q, \infty)$, $q \in \mathbb{Z}$, $q \geq 3$. The modular group $\Gamma_{\mathbb{Z}}$ is given by the signature $(2,3,\infty)$. In general, the triangular groups have been well-examined. In particular, the list of all arithmetic triangular groups is available (see [9]).

The set of arithmetic Fuchsian groups of the first kind (see Section 1) with condition $h \geq 1$ (non-cocompact) is by exact definition the set of groups commensurable with the modular group. One can give a more general, universal definition of an arithmetic Fuchsian group. The group Γ is arithmetic, if it is commensurable with such a group $\Gamma(A, \mathcal{O})$ which is defined as follows. Let K be a totally real field of algebraic numbers of degree n over $\mathbb{Q}$. Let A be a quaternion algebra over K isomorphic over $\mathbb{R}$ to $M_2(\mathbb{R}) \oplus \mathbb{H}^{n-1}$, where $M_2(\mathbb{R})$ is the full matrix algebra of 2×2 matrices, let $\mathbb{H}$ be the Hamilton quaternion algebra. Finally, let $\mathcal{O}$ be the order in A and let

$$U = \{\epsilon \in \mathcal{O} \mid \epsilon\mathcal{O} = \mathcal{O}, \quad n_A(\epsilon) = 1\}$$

be the group of units of $\mathcal{O}$ with the norm n_A which is equal to 1. The group U is embedded in $\mathrm{SL}(2,\mathbb{R})$ and generates the Fuchsian group of the first kind $\Gamma(A, \mathcal{O})$. We note that if $A = M_2(\mathbb{Q})$ and $\mathcal{O}$ is maximal then $\Gamma(A, \mathcal{O}) = \mathrm{PSL}(2,\mathbb{Z})$. If A is a division algebra then the corresponding $\Gamma(A, \mathcal{O})$ is a co-compact Fuchsian group.

We now return to triangular groups, considering first non-cocompact groups. This is the complete Takeuchi's list of arithmetic triangular non-cocompact groups given by their signatures:

$$\begin{gathered}(2,3,\infty),\ (2,4,\infty),\ (2,6,\infty),\ (2,\infty,\infty),\ (3,3,\infty),\\ (3,\infty,\infty),\ (4,4,\infty),\ (6,6,\infty),\ (\infty,\infty,\infty).\end{gathered} \tag{21}$$

We note that all these groups are commensurable with the modular group $(2,3,\infty)$. What is the set of groups commensurable with the triangular non-cocompact group? With regard to this, there is the Margulis theorem which also enables one to characterize the arithmetic groups.

THEOREM 4. *Let Γ be as in Section 1 with condition $h \geq 1$. If Γ is arithmetic then* $\mathrm{Comm}(\Gamma)$ *is a dense set in* $\mathrm{PSL}(2,\mathbb{R})$*; if Γ is non-arithmetic, then* $\mathrm{Comm}(\Gamma)$ *is commensurable with* Γ.

From Theorems 3, 4 it follows that for a non-arithmetic triangular group Δ $\mathrm{Comm}(\Delta)$ is a triangular group. So the set of groups Γ commensurable with non-arithmetic triangular ones is exhausted by the subgroups of finite index of the triangular groups.

The arithmetic non-cocompact groups can also be described up to conjugation as subgroups of finite index of a certain set of maximal groups $\Gamma^*(N)$. Here, the principal result is due to Helling [10]. These groups were examined in detail by Atkin and Lehner, in Kluit (see [11] and bibliography therein).

For $N \in \mathbb{Z}$, $N \geq 1$, we define $\Gamma^*(N)$ as a subgroup of $\mathrm{PSL}(2,\mathbb{R})$ consisting of the matrices

$$m^{-1/2}\begin{pmatrix} a & b \\ c & d \end{pmatrix},$$

where m divides N, N divides c, m divides a and m divides d; $a,b,c,d \in \mathbb{Z}$. We do not check that these conditions define the group. We note that $\Gamma^*(N)$ contains the Hecke congruence subgroup $\Gamma_0(N)$. Recall that

$$\Gamma_0(N) = \left\{\gamma = \begin{pmatrix} a & b \\ c & d \end{pmatrix} \in \mathrm{PSL}(2,\mathbb{Z}) \mid N \quad \text{divides} \quad c\right\}.$$

The group $\Gamma^*(N)$ is the normalizer of $\Gamma_0(N)$. Its quotient group is abelian of type $(2,2,\ldots,2)$, where the number of factors in the direct product is equal to the number of prime divisors of the number N. For almost all N, $\Gamma^*(N)$ is the full normalizer of $\Gamma_0(N)$. Now we present Helling's findings.

THEOREM 5. *Let the group $\Gamma \subset \mathrm{PGL}(2,\mathbb{R})^+$ be commensurable with* $\mathrm{PSL}(2,\mathbb{Z})$*, then there exists a squarefree number N such that Γ is conjugate to a subgroup of* $\Gamma^*(N)$. *(The conjugation is taken in* $\mathrm{PGL}(2,\mathbb{C})$*).*

Thus, the groups $\Gamma^*(N)$ are maximal arithmetic (non-cocompact). The groups in the list (21) are the same. A direct verification shows that for $N = 1,2,3$

$$\begin{aligned}
&(2,3,\infty) \quad \text{is conjugate (coincides) to} \quad \Gamma^*(1),\\
&(2,4,\infty) \quad \text{is conjugate to} \quad \Gamma^*(2),\\
&(2,6,\infty) \quad \text{is conjugate to} \quad \Gamma^*(3).
\end{aligned}$$

In summarizing this small investigation of the groups commensurable with the triangular (non-cocompact) ones, it can be said that it suffices to study the subgroups of finite index of the groups $\Gamma^*(N)$ and non-arithmetic triangular groups.

We now proceed to examining co-compact Fuchsian groups commensurable with triangular ones. In some aspects, this theory is more complex than the previous case. Apparently, the theorem analogous to Theorem 4, takes place, so that non-arithmetic subgroups here are also exhausted by the subgroups of finite index of the triangular ones. As for the set of groups commensurable with the arithmetic triangular (co-compact) groups, we are at present unaware of the relation of this set to the whole set of arithmetic co-compact groups. Takeuchi's complete list of arithmetic co-compact triangular groups (signatures) obtained with the help of the computer, consists of 76 groups (see [9]).

The main theorem, by means of which this list was obtained, is as follows.

THEOREM 6. *Let Γ be a triangular group of the type (m_1, m_2, m_3), $m_1 \neq \infty$, $m_2 \neq \infty$, $m_3 \neq \infty$. Let K_0 be a field such that*

$$K_0 = \mathbb{Q}\big((\cos \pi/m_1)^2, (\cos \pi/m_2)^2, \cos \pi/m_3, (\cos \pi/m_1)(\cos \pi/m_2)(\cos \pi/m_3)\big).$$

The group Γ is arithmetic iff $K_0 = \mathbb{Q}$ or $K_0 \supseteq \mathbb{Q}$ and for any non-identical isomorphism σ of the field K_0 into $\mathbb{R}$ the following inequality is valid:

$$\sigma\big((\cos \pi/m_1)^2 + (\cos \pi/m_2)^2 + (\cos \pi/m_3)^2 + \\ + 2(\cos \pi/m_1)(\cos \pi/m_2)(\cos \pi/m_3) - 1\big) < 0.$$

6. The Schwarz Differential Equation and an Algebraic Automorphic Equation

As is known from the classical theory of automorphic functions, any two analytic functions automorphic relative to a Fuchsian group of the first kind are connected by an algebraic relation. In more detail, let Γ be an arbitrary group from Section 1, f_1, f_2 be two classical meromorphic functions automorphic relative to Γ. Let then f_1 have n_1 poles, and f_2 have n_2 poles in the fundamental domain F of the group Γ. Then there exists an irreducible polynomial $P(x_1, x_2)$ with complex coefficients in two variables, the degree of which by x_1 does not exceed n_2, and by x_2 does not exceed n_1 such that the equality is valid

$$P(f_1, f_2) = 0. \tag{22}$$

If a discrete group Γ has a property $g = 0$, then there exists a univalent function $J(z)$ having only one pole in F. From the previous assertion it follows that for such a group Γ any meromorphic automorphic function f is a rational function in J of which the power of the numerator and the denominator does not exceed the number of poles of f in F.

Equation (22) is called a generalized modular equation or automorphic equation. The relation (22) for $f_1(z) = J(z)$, $f_2(z) = J(Nz)$, where $J(z)$ is the modular invariant (Hauptfunktion) for $\mathrm{PSL}(2, \mathbb{Z})$, $N > 1$ is a natural number, is the classical modular equation. As Γ one takes the Hecke congruence subgroup $\Gamma_0(N)$ mentioned in Section 5.

Now we apply relation (22) to the problem of restoring the Schwarz equation (5) by the group Γ in the case when Γ is commensurable with some triangular group Δ. Here we assume naturally that Γ has $g = 0$. Set $\Gamma \cap \Delta = \Gamma_0$. We denote by J_Γ and J_Δ, respectively, Hauptfunktionen for Γ and Δ. We set $k_1 = [\Gamma : \Gamma_0]$, $k_2 = [\Delta : \Gamma_0]$ to be the indices. As J_Γ and J_Δ are univalent functions on fundamental domains of the groups Γ and Δ, respectively, then considering them as automorphic functions we see that J_Γ has k_1 poles and J_Δ has k_2 poles in the fundamental domain of the group Γ_0. As J_Γ and J_Δ are automorphic meromorphic functions relative to the group Γ_0, they then satisfy the automorphic equation

$$P(J_\Gamma, J_\Delta) = 0, \tag{23}$$

where the degree of the irreducible polynomial in J_Γ (J_Δ) does not exceed k_2 (k_1, respectively). Thus, if the automorphic equation P and Hauptfunktion J_Δ for the triangular group Δ are known, then the problem of defining J_Γ and, therefore, the desired function Q_Γ from (5) reduces to solving the algebraic equation. This idea was widely known to classics for the case of arithmetic groups (see [1], [2] and the excellent survey by Swinnerton–Dyer [12]).

We note that the construction of the function J_Δ is a difficult, but real problem. It reduces to inversing the quotients of hypergeometric functions (19), (20). The explicit solution to the automorphic equation is also a difficult problem, provided the degree is high. It is very important here to have plenty of information on the structure of the corresponding discrete group.

The problem of restoring Equation (5) (the function Q_Γ) by the group Γ can be reduced in many cases only to constructing the automorphic equation, if one makes use of the results in Section 5. In fact, if we put aside those groups Γ which are commensurable with arithmetic triangular co-compact ones and which are not their subgroups of finite index, then the situation when Γ is either a subgroup of the triangular group or a subgroup of $\Gamma^*(N)$ (with $g = 0$) is, apparently, general. Now we consider this case in more detail.

7. General Variable Change in the Schwarz Equation Defined by the Automorphic Equation

Let G be either any triangular group Δ, or the group $\Gamma^*(N)$ with $g = 0$, let $\Gamma \subset G$ be a subgroup of finite index with the same condition $g = 0$, $[G : \Gamma] < \infty$. We denote by J_G the Hauptfunktion for G with some normalization condition. Equation (23) turns into

$$J_G = F(J_\Gamma) \tag{24}$$

where $F(J_\Gamma)$ is the rational function of J_Γ, of which the powers of the numerator and the denominator do no exceed $[G : \Gamma]$. The Schwarz equation (5) for the group

G has the form

$$\{z, J_G\} = Q_G(J_G).$$

We make in it the variable change (24). The following equality can easily be verified:

$$\{z, J_\Gamma\} = \{z, J_G\}F'^2(J_\Gamma) + \{F, J_\Gamma\}.$$

Hence we arrive at

$$\{z, J_\Gamma\} = Q_G(F(J_\Gamma))F'^2(J_\Gamma) + \{F, J_\Gamma\}. \tag{25}$$

Now we use Equation (5). We have

$$Q_\Gamma(J) = Q_G(F(J))F'^2(J) + \{F, J\}. \tag{26}$$

We have proved the following simple but significant theorem.

THEOREM 7. *Let G be a Fuchsian group as in Section 1, let $\Gamma \subset G$ be a subgroup of finite index, the genus of G and Γ is equal to zero. Then the right-hand part of the Schwarz equation $Q_\Gamma(J)$ can be obtained by formula (26), where the rational function F (see (24)) is determined by the automorphic equation for J_G and J_Γ. The powers of numerator and denominator of F do not exceed the index $[G : \Gamma]$.*

Thus, as can be seen from (26), for the definition of Q_Γ it is sufficient to know the functions F and Q_G. We shall discuss later the construction of the automorphic equation (of the function F), the function Q_G is known for the triangular group $G = \Delta$. Let us consider the case

$$G = \Gamma^*(N), \quad g = 0.$$

8. The group $\Gamma^*(N)$ and Its Automorphic Functions

The complete list of the groups $\Gamma^*(N)$ for which $g = 0$ is contained in Kluit [11]. We present this list here consisting of 65 values of N:

$$\begin{aligned} &N \leq 36 \quad \text{and} \\ &N = 38,\ 39,\ 41,\ 42,\ 44,\ 45,\ 46,\ 47,\ 49,\ 50,\ 51,\ 54,\ 55,\ 56,\ 59, \\ &\qquad 60,\ 62,\ 66,\ 69,\ 70,\ 71,\ 78,\ 87,\ 92,\ 94,\ 95,\ 105,\ 110,\ 119. \end{aligned} \tag{27}$$

For each of these groups the calculation of the function Q_G from (26) is possible, although this is an arduous task. The situation can be further simplified if one is restricted (27) only to primes N, i.e.

$$N = 2,\ 3,\ 5,\ 7,\ 11,\ 13,\ 17,\ 19,\ 23,\ 29,\ 31,\ 41,\ 47,\ 59,\ 71. \tag{28}$$

For N from (28), Kluit found Hauptfunktionen J_G in terms of the Dedekind η-function which is the main contribution in computing Q_G. But since Kluit's paper has not yet been published, we shall not give J_G here. We limit ourselves to the values

$$N = 2, 3, 5, 7, 13, \tag{29}$$

for which J_G can be obtained easily. In fact, the following list of values N, for which $\Gamma_0(N)$ has $g = 0$, is widely known:

$$N \leq 10 \quad \text{and} \quad N = 12,\ 13,\ 16,\ 18,\ 25. \tag{30}$$

The set of primes N from (30) coincides exactly with (29). Then we introduce the classical functions

$$\begin{aligned} \eta(z) &= e^{2\pi i z/12} \prod_{n=1}^{\infty} (1 - e^{2\pi i n z}), \\ \Delta(z) &= (2\pi)^{12} \eta^{24}(z), \quad z \in H. \end{aligned} \tag{31}$$

For $N = 2, 3, 5, 7, 13$ we set $r = 24/(N-1)$. The function

$$\Phi(z) = \left(\frac{\eta(Nz)}{\eta(z)} \right)^r$$

is the Hauptfunktion J_{Γ_0} for $\Gamma_0(N)$ (see [14]). For the prime N $\Gamma^*(N)$ differs from $\Gamma_0(N)$ by one involution, roughly speaking, and since for N from (29) both $\Gamma^*(N)$ and $\Gamma_0(N)$ have $g = 0$ then the Hauptfunktion for $\Gamma^*(n)$ is equal to

$$J_G(z) = 1/\Phi(z) + N^{r/2}\Phi(z) = 1/J_{\Gamma_0(z)} + N^{r/2} J_{\Gamma_0}(z). \tag{32}$$

We note that this argument is meaningful at least for $N = 13$ since in this case the group $\Gamma^*(N)$ is not triangular. It is not difficult to verify that for the volumes of fundamental domains for triangular groups from the list (21) the inequality $|F| \leq 2\pi$ is valid, the volume of $|F^*|$ for $\Gamma^*(13)$ is equal to $7\pi/3$.

From formula (32) according to the idea of deriving (26) one obtains the expression for $Q_G(J)$ if we know $Q_{\Gamma_0}(J)$. $Q_{\Gamma_0}(J)$ can be found using $Q_{\Gamma_{\mathbf{z}}}(J)$, since Γ_0 is a subgroup of the modular group, but for this aim it is necessary to construct a modular equation that links the invariants J_{Γ_0} and $J_{\Gamma_{\mathbf{z}}}$ which will be done later on.

Thus, the main question we must now address is that of constructing the automorphic equation (24). But it is first useful to gain some general information on subgroups of finite index (in particular, with condition $g = 0$) of the triangular groups, and the groups $\Gamma^*(N)$ with condition $g = 0$ as well.

9. Subgroups of Finite Index of Fuchsian Groups

As can be seen from the above, we are now interested in the groups Γ with signature $(0, n_1, \ldots, n_k, h)$ and their subgroups of finite index with signatures of the same type. The groups Γ with signature $(0, n_1, \ldots, n_k, 0$ are studied less often since they are not free products. So first we shall assume $h \geq 1$.

The Kurosh theorem (see [15]) is the significant theorem describing the structure of many Fuchsian groups from the standpoint of general group theory.

THEOREM 8. *Let Γ be a free product of subgroups A_α, i.e. $\Gamma = \prod^*_\alpha A_\alpha$. Let $\overset{\circ}{\Gamma} \subset \Gamma$ be its subgroup. Then $\overset{\circ}{\Gamma}$ is also a free product of the following form: $\overset{\circ}{\Gamma} = F * \prod^*_\beta B_\beta$, where F is a free group, and each group B_β is conjugate in Γ to some subgroup of A_α.*

How many subgroups are there in a given Fuchsian group? More precisely, one must characterize the function $M_\Gamma(N)$, that is the number of subgroups of index N in Γ. Under our assumptions $g = 0$, $h \geq 1$ this can be done (see [16], [17]). Γ is the free product of finitely many cyclic groups $\Gamma = C_{m_1} * C_{m_2} * \cdots * C_{m_n}$, C_m being the cyclic group of order $m \leq \infty$. $M_\Gamma(N)$ can be expressed in terms of functions $\tau_m(N)$, where $\tau_m(N)$ is the number of homomorphisms of the group C_m into the symmetric group S_N. We set

$$\tau_{m_1}(N)\tau_{m_2}(N)\cdots\tau_{m_n}(N)/N! = \alpha_N.$$

Then $M_\Gamma(N)$ satisfies the recurrent relation

$$\sum_{l=1}^{N} \alpha_{N-l} M_\Gamma(l) = N \cdot \alpha_N \quad N \geq 1, \quad \alpha_0 = 1. \tag{33}$$

The functions $\tau_m(N)$ are well-examined. It is clear that $\tau_\infty(N) = N!$ and $\tau_m(N)$ is the number of elements in S_N with the orders that divide m. A generating function for $\tau_m(N)$ is equal to

$$\sum_{N=0}^{\infty} \frac{\tau_m(N)}{N!} x^N = \exp\Big(\sum_{d|m} \frac{x^d}{d}\Big),$$

which singles out explicitly the prime values of m. From (33) it follows the main asymptotic formula

$$M_\Gamma(N) \sim \tau_{m_1}(N)\tau_{m_2}(N)\cdots\tau_{m_n}(N)/(N-1)! \tag{34}$$

For Γ with the prime orders of the cyclic groups C_m there exists an elegant asymptotic formula for α_N that is connected with the volume of the fundamental domain F of Γ (3). So, if we keep the notation (8) and assume that all n_j are primes then it emerges [17] that

$$\alpha_N \underset{N\to\infty}{\sim} K \exp\Big[\frac{\mu(F)}{2\pi} N \ln N - \frac{\mu(F)}{2\pi} N + \sum_{j=1}^{k} N^{1/n_j} + \big(\tfrac{h}{2} - 1\big) \ln N\Big], \tag{35}$$

$$\mu(F) = |F|$$

i.e., the function $M_\Gamma(N)$ increases very quickly and the reserve of subgroups of finite index in Γ is sufficiently large.

But if we set the problem of examining the function $M_\Gamma(N)$ assuming the other parameters g, h and so on of the subgroup (see (3)) to be given (for example, it is especially interesting when $g = 0$), then the answer is much more complicated, and in general it is difficult to prove the existence of at least one subgroup with given properties. Now we shall discuss this question in more detail, but first we note that the asymptotic formula is known for $M_\Gamma(N)$ similar to (34) with condition: the considered subgroups of index N are free (see [18]).

Now let us give an arbitrary group Γ with signature $(g, n_1, \ldots, n_k, h)$ (see Section 1) and its subgroup $\mathring{\Gamma}$ of finite index $d = [\Gamma : \mathring{\Gamma}]$ with signature $(\mathring{g}, \mathring{n}_1, \ldots, \mathring{n}_{\mathring{k}}, h)$. It is widely known that these two signatures are linked by the following Riemann–Hurwitz relation

$$2\mathring{g} - 2 + \mathring{h} + \sum_{j=1}^{\mathring{k}} (1 - 1/\mathring{n}_j) = d\Big[2g - 2 + h + \sum_{j=1}^{k} (1 - 1/n_j)\Big], \tag{36}$$

which is equivalent thanks to the Gauss–Bonnet formula (3) to the equality $|\mathring{F}| = d|F|$, where $\mathring{F}$ is the fundamental domain of the group $\mathring{\Gamma}$. In addition to (36) the inequalities

$$g \leq \mathring{g}, \quad h \leq \mathring{h} \leq dh \tag{37}$$

and the division conditions: each $\mathring{n}_j$ divides some n_i, are fulfilled.

Let there correspond to a finite cyclic subgroup A_α of order n_α r_α subgroups $B_\beta(\alpha)$ with orders $\mathring{n}_\beta(\alpha)$, where, evidently, $\mathring{n}_\beta(\alpha) \mid n_\alpha$, $1 \leq \beta \leq r_\alpha$. In this new notation

$$(\mathring{g}, \mathring{n}_1, \ldots, \mathring{n}_{\mathring{k}}, \mathring{h}) = (\mathring{g}, \mathring{n}_1(1), \mathring{n}_2(1), \ldots, \mathring{n}_{r_1}(1), \ldots, \mathring{n}_1(k), \mathring{n}_2(k), \ldots, \mathring{n}_{r_k}(k), \mathring{h}). \tag{38}$$

There are two approaches to proving the existence of the subgroup with signature (38) of a given Fuchsian group. The first reduces the problem to proving the existence of special permutations in the symmetric group and to finding them (Millington [19]), Singerman [20]). The second method is based on topological examination of the problem (Kulkarni [21], [22]).

These methods allow one to investigate many groups, but the general situation remains as yet unsolved. As indicated in [21], the problem relates to the Hurwitz problem of realizability of special coverings.

We now give Singerman's theorem providing the necessary and sufficient conditions for the existence of the subgroup of Γ in terms of permutations. The assertion is a generalization of a remarkable Millington theorem that succeeds in proving the existence of desired permutations for the modular group.

THEOREM 9 ([20]). *In order that in* Γ *there exists a subgroup* $\mathring{\Gamma}$ *of index* d *with signature* (38) *it is necessary and sufficient that the following conditions be fulfilled:* (1) *there exists the finite group of permutations* G *which is transitive on* d *points and the epimorphism* $\theta : \Gamma \to G$ *such that the permutation* $\theta(E_j)$ (*see Section* 1) *has exactly* r_j *cycles of lengths less than* n_j; *the lengths of these cycles are* $n_j/n_\alpha(j)$, $1 \leq \alpha \leq r_j$. *In addition, the following equality holds:*

$$\mathring{h} = \sum_{k=1}^{h} \delta(V_k)$$

where $\delta(\gamma)$ *is the number of cycles in the permutation* $\theta(\gamma)$; (2) *formula* (36) *is valid.*

Topological methods enable one to obtain the more effective necessary and sufficient conditions for the existence of a subgroup with given invariants. Kulkarni showed that in the case of free product the solvability of a certain Diophantine system of equations should be added to necessary conditions (36), (37) in order that the conditions thus obtained be sufficient. We now give this system. Let $\mathring{n}_1, \ldots, \mathring{n}_l$ be a maximal set of distinct $\mathring{n}_j$ and each $\mathring{n}_q$ enters b_q times, $1 \leq q \leq l$. We set

$$\epsilon_{iq} = \begin{cases} 0, & \mathring{n}_q \nmid n_i \\ 1, & \mathring{n}_q \mid n_i, \end{cases} \qquad \delta_{iq} = n_i/\mathring{n}_q.$$

Then the solvability of the system

$$\exists\, x_{iq} \in \mathbb{Z}_+ \Longrightarrow \begin{cases} \sum_i \epsilon_{iq} x_{iq} = b_q \\ \sum_q \delta_{iq} x_{iq} = d \end{cases} \tag{39}$$

is the additional condition to (36), (37).

THEOREM 10 ([22]). *The conditions* (36), (37), (39) *are necessary and sufficient for existence of the subgroup* $\mathring{\Gamma}$ *of* Γ.

As will be seen later, the conditions (39) are more easily verifiable than the conditions of Theorem 9.

10. Subgroups of Finite Index of the Triangular Groups

For the triangular groups Δ and their subgroups $\mathring{\Delta}$ of finite index it is useful to write out the Riemann–Hurwitz relation in detail. We restrict ourselves here to consideration of the most simple situation when the formulas are easily visible and the result is effective. The question is of the special triangular groups, i.e. the free products with prime signatures. We consider three types of such signatures:

(1) (∞,∞,∞), (2) (m_1,∞,∞), $m_1 \geq 2$, (3) (m_1,m_2,∞), $2 \leq m_1 \leq m_2 < \infty$; m_1, m_2 are primes.

(1) Relation (36) for $\overset{\circ}{\Delta}(=\overset{\circ}{\Gamma})$ is equal to

$$2\overset{\circ}{g} - 2 + \overset{\circ}{h} = d. \tag{40}$$

We recall that in this case Δ is generated by three parabolic generators V_1, V_2, V_3 with relation $V_1V_2V_3 = 1$ (see Section 1). Each primitive parabolic conjugacy class in $\Delta\{V_i\}$, $i = 1,2,3$, splits in $\overset{\circ}{\Delta}$ into h_i primitive classes. The generators of corresponding parabolic cyclic groups are conjugate in Δ to $V_i^{\mu_{ij}}$, $1 \leq i \leq 3$, $1 \leq j \leq h_i$. For the widths of cusps μ_{ij} the natural relation is valid

$$\sum_{i=1}^{3}\sum_{j=1}^{h_i} \mu_{ij} = 3d, \quad h_1 + h_2 + h_3 = \overset{\circ}{h}, \tag{41}$$

together with (40) characterizing the subgroup $\overset{\circ}{\Delta} \subset \Delta$.

(2) The group Δ is generated by E, V_1, V_2 with relations $EV_1V_2 = 1$, $E^m = 1$. (Recall that $m \geq 2$, m is prime). Each primitive parabolic class in $\Delta\{V_i\}$, $i = 1,2$, splits in $\overset{\circ}{\Delta}$ into h_i primitive classes. The generators of corresponding parabolic groups are conjugate in Δ to $V_i^{\mu_{ij}}$, $1 \leq i \leq 2$, $1 \leq j \leq h_i$. For the widths of cusps μ_{ij} the following relation takes place

$$\sum_{i=1}^{2}\sum_{j=1}^{h_i} \mu_{ij} = 2d, \quad h_1 + h_2 = \overset{\circ}{h}. \tag{42}$$

Primitive elliptic classes $\{E\}$ splits in $\overset{\circ}{\Delta}$ into e primitive classes of the same order. So the Riemann–Hurwitz relation takes the form

$$2\overset{\circ}{g} - 2 + \overset{\circ}{h} = (d - e)(1 - 1/m). \tag{43}$$

(3) The group Δ is generated by E_1, E_2, V with the relations $E_1E_2V = 1$, $E_1^{m_1} = E_2^{m_2} = 1$, m_1, m_2 are primes ($2 \leq m_1 \leq m_2 < \infty$). Primitive parabolic class in Δ splits in $\overset{\circ}{\Delta}$ into $\overset{\circ}{h}$ primitive classes. The generators of corresponding parabolic cyclic groups are conjugate in Δ to V^{μ_i}, $1 \leq i \leq \overset{\circ}{h}$. For the widths of cusps μ_i the following relation is fulfilled

$$\mu_1 + \mu_2 + \cdots + \mu_{\overset{\circ}{h}} = d. \tag{44}$$

Primitive elliptic class $\{E_i\}$, $i = 1,2$, splits in $\overset{\circ}{\Delta}$ into e_i primitive classes of the same order. The relation (36) takes the form

$$2\overset{\circ}{g} - 2 + \overset{\circ}{h} + e_1(1 - 1/m_1) + e_2(1 - 1/m_2) = d(1 - 1/m_1 - 1/m_2). \tag{45}$$

For the subgroups of the triangular groups Δ considered above, Theorems 9 and 10 turn out to be effective. The most simple is the verification of solvability of Diophantine system (39) which proves to be solvable. We state the more precise result.

THEOREM 11. *Let Δ have signature (p, q, ∞) where p and q are distinct primes. In order that there exists a subgroup $\overset{\circ}{\Delta}$ in Δ of finite index d, of genus $\overset{\circ}{g} = 0$, with other parameters $\overset{\circ}{h} \geq 1$, $e_1 \geq 0$, $e_2 \geq 0$ it is necessary and sufficient only that the Riemann–Hurwitz relation (45) be fulfilled.*

We now proceed to the case of the subgroups of the modular group that is studied most often.

11. Subgroups of Finite Index of the Modular Group

First of all, we wish to make more precise the results of Sections 8, 10 in the case of the modular group $\Gamma_{\mathbf{Z}} = C_2 * C_3$. For the asymptotics of the function $M(N)$ the following formula is valid

$$M(N) \sim \tau_2(N)\tau_3(N)/(N-1)! \sim$$
$$\sim \left(12\pi e^{1/2}\right)^{-1/2} \exp\left[\tfrac{N}{6}\ln N - \tfrac{N}{6} + N^{1/2} + N^{1/3} + \tfrac{1}{2}\ln N\right], \quad N \to \infty,$$

i.e., it increases sufficiently rapidly. There are several values of $M(N)$ obtained by means of relation (33)

N	1	2	3	4	5	6	7	8	9	10
$M(N)$	1	1	4	8	5	22	42	40	120	265

(46)

The table of $M(N)$ up to $N = 100$ is given in [17].

The Riemann–Hurwitz relation for $\Gamma_{\mathbf{Z}}$ takes the form

$$d = 12\overset{\circ}{g} - 12 + 6\overset{\circ}{h} + 3e_1 + 4e_2 \quad (\text{in } (45) \quad m_1 = 2,\ m_2 = 3). \tag{47}$$

The formula for the width of cusps (44) is the same

$$\mu_1 + \mu_2 + \cdots + \mu_{\overset{\circ}{h}} = d. \tag{48}$$

According to Millington, formula (47) is the necessary and sufficient condition for existence of the subgroup $\overset{\circ}{\Gamma}$ with parameters $\overset{\circ}{g}$, $\overset{\circ}{h}$, e_1, e_2, d.

However, for the modular group the aforementioned results can be improved on. It was known in the last century that there exist infinitely many subgroups of finite index in $\Gamma_{\mathbf{Z}}$ with $\overset{\circ}{g} = 0$. Then this result was generalized by Jones and Singerman by means of the combinatorial theory of the maps of oriented surfaces to any genus $\overset{\circ}{g} \geq 0$ (see [23]). For the subgroup with the number of parabolic classes $\overset{\circ}{h}$ equal to 1 or 2 and $\overset{\circ}{g} = 0$ Petersson [24] obtained essentially more effective results. Namely, let $M(d, \overset{\circ}{g}, \overset{\circ}{h})$ be the number of subgroups of $\Gamma_{\mathbf{Z}}$ of index d, genus $\overset{\circ}{g}$ and with the number of primitive parabolic classes $\overset{\circ}{h}$. The Petersson theorem is as follows.

THEOREM 12. *For* $\overset{\circ}{g} = 0$, $\overset{\circ}{h} = 1$ *or* 2 *there exist the constants* $K(0,1)$, $K(0,2)$ *and* $c \in (1,2)$ *such that*

$$M(d, \overset{\circ}{g}, \overset{\circ}{h}) \geq K(\overset{\circ}{g}, \overset{\circ}{h}) c^d. \tag{49}$$

By means of more refined techniques of diagrams Stothers improved Petersson's results for more general situations (see [25], [26]). We give here the Stothers theorem only in the case $\overset{\circ}{g} = 0$.

THEOREM 13. *The following asymptotic formula is valid*

$$M(d, 0, \overset{\circ}{h}) = M(6\overset{\circ}{h} - 12, 0, \overset{\circ}{h}) K_1 d^{3(\overset{\circ}{h}/2-1)} c^d (1 + O(1/\alpha)), \quad d \to \infty, \tag{50}$$

with a certain constant K_1. *The absolute constant* c *is one of the roots of the equation*

$$x^4 = 4(x+1), \quad c \approx 1,8.$$

Up to this point we have discussed the subgroups with given main topological characteristics $\overset{\circ}{g}$, $\overset{\circ}{h}$, e_1, e_2 and, of course, d or, as one says, with short specification $(\overset{\circ}{g}, d, \overset{\circ}{h}, e_1, e_2)$ to be given. The set of numbers $(\overset{\circ}{g}, d, \overset{\circ}{h}, e_1, e_2, \mu_1, \ldots, \mu_{\overset{\circ}{h}})$, i.e. a short specification plus the set of cusp widths, is the complete specification of the subgroup $\overset{\circ}{\Gamma}$. As mentioned above, the Millington theorem asserts that the Riemann–Hurwitz relation (47) is the necessary and sufficient condition for the existence of the subgroup Γ with corresponding short specification. But there are well-known examples of sets of natural numbers μ_j satisfying condition (48) that correspond to no subgroup $\overset{\circ}{\Gamma}$ with the corresponding complete specification. The first examples, as far as we know, of 'impossible' specifications were found by Petersson and Millington.

As in Theorem 13, using his diagram technique, simplifying Millington's proof of the theorem mentioned above, Stothers found the domain of parameters d, $\overset{\circ}{h}$ in which there are no impossible specifications. More precisely, the theorem of Stothers is as follows [27].

THEOREM 14. *For any* $\epsilon > 0$ *there exists an effective constant* $N(\epsilon)$ *such that if* $d \geq (6+\epsilon)\overset{\circ}{h} + N(\epsilon)$, *then any set of natural numbers* μ_j *with the condition* (48) *corresponds to the subgroup* $\overset{\circ}{\Gamma}$.

This short survey of the theory of subgroups of the modular group remains unsatisfactory unless certain significant classical subgroups, and more arithmetic classification of the subgroups $\overset{\circ}{\Gamma}$, are also referred to.

The principal congruence subgroup $\Gamma(m)$ is defined as usual

$$\Gamma(m) = \{\gamma \in \Gamma_{\mathbf{Z}} \mid \gamma \equiv \pm 1 (\operatorname{mod} m)\}, \quad m > 1.$$

Its index $d = d(m)$ is equal to

$$d(m) = [\Gamma_{\mathbf{Z}} : \Gamma(m)] = \begin{cases} 6, & m = 2 \\ m^{3/2} \prod\limits_{p \mid m} (1 - 1/p^2), & m > 2, \end{cases}$$

where the product is taken over prime divisors. $\Gamma(m)$ has no elements of finite order $\neq 1$, so (47) takes the form

$$d(m) = 12\overset{\circ}{g}(m) - 12 + 6\overset{\circ}{h}(m).$$

In addition, $\Gamma(m)$ is the normal subgroup, which implies that all cusps have the same widths (m, in this case). Formula (48) turns into

$$\overset{\circ}{h}(m) \cdot m = d(m).$$

Thus, the genus $\overset{\circ}{g}(m)$ can be calculated by the formula

$$\overset{\circ}{g}(m) = 1 + \delta \frac{m^2}{24}(m - 6) \prod_{p \mid m} (1 - 1/p^2), \quad \delta = \begin{cases} 2, & m = 2 \\ 1, & m > 2 \end{cases}$$

and this genus is equal to zero only for the subgroups $\Gamma(2)$, $\Gamma(3)$, $\Gamma(4)$ and $\Gamma(5)$. The corresponding index $d(m)$ is equal to 6, 12, 24, 60. We note that $\Gamma(2)$ is the triangular group with signature (∞, ∞, ∞).

The indicated groups $\Gamma(2) - \Gamma(5)$ were well covered by classics since the corresponding quotients $\Gamma_{\mathbf{Z}}/\Gamma(n)$, $n = 2, \ldots, 5$, are isomorphic to symmetry groups for regular polyhedrons.

We have defined the Hecke congruence subgroup $\Gamma_0(m)$ and we have considered its properties. The main invariants of this group can be computed and may be presented by the formulas

$$d = m \prod_{p \mid m} (1 + 1/p), \quad e_1 = \begin{cases} 0, & 4 \mid m \\ \prod\limits_{p \mid m} (1 + (\frac{-1}{p})), & 4 \nmid m, \end{cases}$$

$$e_2 = \begin{cases} 0, & 9 \mid m, \\ \prod\limits_{p \mid m} (1 + (\frac{-3}{p})), & 9 \nmid m, \end{cases} \quad \overset{\circ}{h} = \sum_{\substack{d \mid m \\ d \geq 0}} \varphi((d, \tfrac{m}{d})),$$

where φ is the Euler function and for the Legendre symbol we have

$$\left(\frac{-1}{p}\right) = \begin{cases} 0, & p = 2 \\ 1, & p \equiv 1 \operatorname{mod} 4 \\ -1, & p \equiv 3 \operatorname{mod} 4 \end{cases} \quad \left(\frac{-3}{p}\right) = \begin{cases} 0, & p = 3 \\ 1, & p \equiv 1 \operatorname{mod} 3. \\ -1, & p \equiv 2 \operatorname{mod} 3 \end{cases}$$

For prime values of $m = p$ these formulas can be simplified

$$d = p + 1, \quad e_1 = 1 + \left(\tfrac{-1}{p}\right), \quad e_2 = 1 + \left(\tfrac{-3}{p}\right), \quad \overset{\circ}{h} = 2.$$

As can be seen from the Riemann–Hurwitz formula for such values of $m = p$, $\Gamma_0(p)$ has the zero genus only for $p = 2, 3, 5, 7, 13$.

There exist different methods of classification of the subgroups of finite index in $\Gamma_{\mathbf{Z}}$. The normal and congruence subgroups are important classes of such subgroups. It is evident that any subgroup $\overset{\circ}{\Gamma}$ of index $d < \infty$ in $\Gamma_{\mathbf{Z}}$ contains a normal subgroup of finite index in $\Gamma_{\mathbf{Z}}$ as well. As such a subgroup, one can take the intersection of all subgroups conjugate to $\overset{\circ}{\Gamma}$ in $\Gamma_{\mathbf{Z}}$.

Normal subgroups of finite index in $\Gamma_{\mathbf{Z}}$ are well recognized (see [28], [29]). With a few exceptions, any such subgroup is free, i.e. in this case it does not contain elements of finite order. These exceptions can be described thus. Let $\Gamma_{\mathbf{Z}}^m \subset \Gamma_{\mathbf{Z}}$ be a subgroup generated by m-th powers of elements $\gamma \in \Gamma_{\mathbf{Z}}$. The indicated exceptions (besides $\Gamma_{\mathbf{Z}}$ itself) are $\Gamma_{\mathbf{Z}}^2$, $\Gamma_{\mathbf{Z}}^3$.

Thus, for the normal subgroup of index d distinct from $\Gamma_{\mathbf{Z}}$, $\Gamma_{\mathbf{Z}}^2$, $\Gamma_{\mathbf{Z}}^3$ the Riemann–Hurwitz formula (47) takes the form

$$\mathrm{d} = 12\overset{\circ}{g} - 12 + 6\overset{\circ}{h}, \tag{51}$$

whence it can be seen that its index must be divided by 6. In addition, for such a subgroup the widths of cusps are the same and by virtue of formula (48) equal to some natural number

$$n = d/\overset{\circ}{h}.$$

Formula (51) turns into

$$\overset{\circ}{g} = 1 + \mathrm{d}\,\frac{n-6}{12n}. \tag{52}$$

It can be seen from (52) that for fixed $\overset{\circ}{g} \neq 1$ there exists only a finite number of normal subgroups of the group $\Gamma_{\mathbf{Z}}$ of finite index and given genus $\overset{\circ}{g}$. In particular, for $\overset{\circ}{g} = 0$ all normal subgroups of $\Gamma_{\mathbf{Z}}$ of finite index are given by the list

$$\Gamma_{\mathbf{Z}}, \Gamma_{\mathbf{Z}}^2, \Gamma_{\mathbf{Z}}^3, \Gamma(2), \Gamma(3), \Gamma(4), \Gamma(5), \tag{53}$$

where $\Gamma(m)$ is the principal congruence subgroup of $\Gamma_{\mathbf{Z}}$ as before.

As subgroup $\overset{\circ}{\Gamma} \subset \Gamma_{\mathbf{Z}}$ containing for some m the principal congruence subgroup $\Gamma(m)$ is called a congruence subgroup. It is called a congruence subgroup of level m, if m is the smallest natural number with such a property. The congruence subgroups are characterized by a finite number of polynomial congruences on matrix elements and for them there exists the rich arithmetical theory of modular forms and functions with various applications.

We note that the normal congruence subgroups allow a rather detailed classification (see [30]).

The fact that there exist non-congruence subgroups of $\Gamma_{\mathbb{Z}}$ was known long ago (see [31], [32]). Current proof of this fact consists of examining the completions of the group $\Gamma_{\mathbb{Z}}$ by two topologies. In one of them, congruence subgroups are neighbourhoods of unit, in the other the role of neighbourhoods of unit play all subgroups of finite index. These two completions lead to totally different objects.

There are known examples of infinite series of normal non-congruence subgroups (with $\overset{\circ}{g} = 1$, naturally). We present here such an example which is attributed to Newman (see [33]).

We recall that the modular group is the free product of two cyclic groups of order 2 or 3. The corresponding generators are

$$T_z = -1/z, \quad P_z = -1/(z+1), \quad P^3 = T^2 = 1. \tag{54}$$

Let Γ' be the commutant of $\Gamma_{\mathbb{Z}}$. This is a free subgroup of rank 2, index 6 in $\Gamma_{\mathbb{Z}}$ which is generated by the elements

$$A = TPTP^2, \quad B = TP^2TP. \tag{55}$$

We introduce the functions $e_A(\gamma)$, $e_B(\gamma)$ which mean the sum of exponents in the word $\gamma \in \Gamma'$ for the generators A and B respectively.

We define the subgroup $\Gamma'' \subset \Gamma'$ by a condition

$$\Gamma'' = \{\gamma \in \Gamma' \mid e_A(\gamma) = e_B(\gamma) = 0\}.$$

The following are necessary and sufficient conditions for the existence in $\Gamma_{\mathbb{Z}}$ of a normal subgroup $\overset{\circ}{\Gamma}$ of finite index with $\overset{\circ}{g} = 1$:

(1) $\Gamma' \supset \overset{\circ}{\Gamma} \supset \Gamma''$;

(2) There exist $p, m, d \in \mathbb{Z}$, $p > 0$, $0 \leq m \leq d-1$, $m^2 + m + 1 \equiv 0 (\mathrm{mod}\, d)$ such that

$$\overset{\circ}{\Gamma} = \{\gamma \in \Gamma' \mid e_A(\gamma) \equiv 0 (\mathrm{mod}\, p), \; e_B(\gamma) \equiv 0 (\mathrm{mod}\, dp)\}. \tag{56}$$

We have defined the notion of level for congruence subgroups in $\Gamma_{\mathbb{Z}}$. Wohlfahrt extended this notion to an arbitrary subgroup of finite index in $\Gamma_{\mathbb{Z}}$ (see [34]). Namely, the least common multiple of widths μ_j of all cusps, $1 \leq j \leq \overset{\circ}{h}$, is the level of $\overset{\circ}{\Gamma}$. These two definitions coincide for the congruence subgroups.

Now we return to the group $\overset{\circ}{\Gamma}$ from (56). This group can be characterized by three numbers $\{p, m, d\}$. It may be shown from the definition that its level is equal to 6. Consequently, $\overset{\circ}{\Gamma}$ is a congruence subgroup in $\Gamma_{\mathbb{Z}}$ iff $\overset{\circ}{\Gamma} \supset \Gamma(6)$. But the index of $\overset{\circ}{\Gamma}$ in Γ' is equal to dp^2, and the index of $\Gamma(6)$ in Γ' is equal to 12, therefore $dp^2 \mid 12$. Hence it follows that only for four values of triples $\{p, m, d\}$ the group $\overset{\circ}{\Gamma}$

is the congruence subgroup:

$$\{p, m, d\} = \{1, 0, 1\},\ \{1, 1, 3\},\ \{2, 0, 1\},\ \{2, 1, 3\},$$

and we have an infinite series of normal non-congruence subgroups of genus 1.

In connection with the differential equations under consideration, we are especially interested in subgroups $\overset{\circ}{\Gamma} \subset \Gamma_{\mathbf{Z}}$ of genus zero. As can be seen from (53) all normal subgroups of finite index and genus zero in $\Gamma_{\mathbf{Z}}$ are simultaneously the congruence subgroups of $\Gamma_{\mathbf{Z}}$. For this it is sufficient to verify that $\Gamma_{\mathbf{Z}}^2$ contains $\Gamma(2)$ with index 3 and $\Gamma_{\mathbf{Z}}^3$ contains $\Gamma(3)$ with index 4. Their levels are equal to 2 and 3 respectively.

As for arbitrary congruence subgroups of finite index in $\Gamma_{\mathbf{Z}}$ of genus zero, there is here an interesting and unproved Rademacher conjecture* that there are only a finite number of such subgroups (see [35]). In connection with this conjecture, two theorems should be mentioned.

THEOREM 15 (see [35]). *A free congruence subgroup of* $\Gamma_{\mathbf{Z}}$ *of the level relatively prime to* $2 \cdot 3 \cdot 4 \cdot 7 \cdot 13$ *has necessarily a strictly positive genus.*

THEOREM 16 (see [36]). *The Rademacher conjecture is valid for the congruence subgroups of prime level.*

Concluding this section, we give an important, classical example of infinite series of subgroups in $\Gamma_{\mathbf{Z}}$ of genus zero. All subgroups with the exception of a finite number of them are non-congruence subgroups (see [31], [32], [37], [38]). This family of groups is the set of normal subgroups of the principal congruence subgroup $\Gamma(2) \subset \Gamma_{\mathbf{Z}}$ (but non-normal in $\Gamma_{\mathbf{Z}}$, in general). The group $\Gamma(2)$ is free and can be given by two free parabolic generators

$$X_z = TPTP_z = z + 2, \quad Y_z = TP^{-1}TP_z^{-1} = \frac{z}{2z+1}, \quad z \in H. \tag{57}$$

Then the desired family of subgroups $\overset{\circ}{\Gamma}_n \subset \Gamma(2) \subset \Gamma_{\mathbf{Z}}$ can be given by $n+1$ free parabolic generators

$$X^n,\ X^rYX^{-r} \quad (0 \leq r \leq n-1). \tag{58}$$

These groups can be defined differently

$$\overset{\circ}{\Gamma}_n = \{\gamma \in \Gamma(2) \mid e_x(\gamma) \equiv 0 (\operatorname{mod} n)\} \tag{59}$$

(see the notation in (56)). The index d_n of the group $\overset{\circ}{\Gamma}_n$ in $\Gamma_{\mathbf{Z}}$ is equal to $6n$. Its genus $\overset{\circ}{g}_n$ is equal to zero since the group is defined only by parabolic generators.

* As P. Zograf has informed me, an effective proof of the Rademacher conjecture easily follows from the results of [46].

It is clear that the level of the subgroup $\overset{\circ}{\Gamma}_n$ in $\Gamma_{\mathbf{Z}}$ is equal to $2n$. So the necessary and sufficient condition for $\overset{\circ}{\Gamma}_n$ to be a congruence subgroup is the condition

$$\overset{\circ}{\Gamma}_n \supset \Gamma(2n). \tag{60}$$

The verification of (60) shows that only for $n = 1, 2, 4, 8$ is the group $\overset{\circ}{\Gamma}_n$ a congruence subgroup in $\Gamma_{\mathbf{Z}}$.

We conclude this section with a theorem generalizing the example given above.

THEOREM 17 (see [39]). *For any integer $g \geq 0$ there exist infinitely many non-congruence subgroups $\overset{\circ}{\Gamma} \subset \Gamma_{\mathbf{Z}}$ of genus $\overset{\circ}{g} = g$.*

12. The Main Theorem on the Automorphic Equation

Let Γ be an arbitrary Fuchsian group from Section 1 with $g = 0$, let F be its fundamental domain. Following the notation of formula (8), let $\{b_j\}$, $1 \leq j \leq n$ be the set of vertices of F. We denote by $J(z)$ the Hauptfunktion for the group Γ normalized by conditions

$$a_1 = 0, \quad a_2 = 1, \quad a_n = \infty,$$

where $a_j = J(b_j)$, $1 \leq j \leq n$. We give here the useful theorem characterizing the main automorphic equation (22).

THEOREM 18. *In order that the function $f(z)$ be an automorphic one-to-one analytic function outside the poles for a subgroup $\overset{\circ}{\Gamma} \subset \Gamma$ of finite index d, it is necessary for f to be an algebraic function J with its only branch points at $J(b_j) = a_j$, $1 \leq j \leq n$. In addition, the order of branching at the point a_j must divide the number m_j (see (8)).*

The proof of this theorem can be outlined as follows. This is a simple generalization to the classical Klein–Fricke theorem (see [40]) for the modular group. Let f be a one-to-one analytic function automorphic relative to $\overset{\circ}{\Gamma}$ and let this be the maximal subgroup of Γ for which f is automorphic. The function $J(z)$ is, in particular, automorphic relative to $\overset{\circ}{\Gamma}$. Consequently, two meromorphic functions are given on the Riemann surface $\widehat{\overset{\circ}{F}}$, and in addition, J takes each value d times. So there exists an irreducible polynomial

$$a_0(J) f^d + a_1(J) f^{d-1} + \cdots + a_d(J) = 0, \tag{61}$$

where $a_j(J)$ are polynomials depending on f. Thus, f is a d-valued algebraic function. Let further (f_0, J_0) be a branch point of a given algebraic function. One can assume that $f_0 < \infty$, otherwise we take $1/f_0$. Suppose that the order of branching is equal to m. Then, locally, in a neighbourhood of the branch point of algebraic function, f expands into the series of powers $(J - J_0)^{1/m}$.

On the other hand, let $z_0 \in H/\overset{\circ}{\Gamma}$ and $J_0 = J(z_0)$. Consider the set of points

$$\Omega_j = \{\gamma b_j \mid \gamma \in \Gamma/\overset{\circ}{\Gamma}\}.$$

Assume that z_0 lies outside the set

$$\Omega = \bigcup_{j=1}^{n} \Omega_j;$$

then in a neighbourhood of z_0 $J(z)$ expands into the power series

$$J(z) = \sum_{n=0}^{\infty} c_n (z - z_0)^n$$

and locally $J - J_0 \sim c_1(z - z_0)$. So for such J_0 $m = 1$, or else we arrive at a contradiction with singlevaluedness of f.

Consequently, the branch point J_0 can correspond only to the point z_0 lying in the set Ω. Now suppose that $z_0 \in \Omega_j$, $j \neq n$. Then in a neighbourhood of z_0 $J(z_0)$ expands into the power series

$$J(z) = \sum_{n=0}^{\infty} \widetilde{c}_n (z - z_0)^{nm_j}$$

and locally $J - J_0 \sim \widetilde{c}_1(z - z_0)^{m_j}$. Since $f(z)$ is $1 - 1$ in the neighbourhood of z_0, then m divides m_j.

Consider now the case when $z_0 \in \Omega_j$, $j \neq n$, $m_j = \infty$, i.e. the corresponding vertex is a cusp. Then $J(z)$ expands in the neighbourhood of $z = z_0$ into the Fourier series

$$f(g_j \widetilde{z}) = \sum_{n=0}^{\infty} \hat{c}_n e^{2\pi i n \widetilde{z}}.$$

(We transfer the cusp at ∞ and make the width of the cusp to be equal to 1.) The existence at $J = J_0$ of the branch point does not contradict the singlevaluedness of the function. The case $z_0 \in \Omega_n$, $m_n = \infty (m_n < \infty)$ can be examined by analogy. Here one should consider $\widetilde{J} = 1/J$ instead of J. The proof is complete.

For a large class of groups Γ one can prove that the necessary conditions of Theorem 18 are also sufficient. For the modular group it was proved by Klein and Fricke and was updated by Atkin and Swinnerton-Dyer. Their proof can be carried over verbatim to the case of triangular groups from Theorem 11. The sufficiency of these conditions for the modular group, for example, is the important assertion for the theory of automorphic forms relative to non-congruence subgroups since it gives an effective tool (practically unique) of constructing both the forms and the subgroups $\overset{\circ}{\Gamma}$ themselves.

13. 'J-Equations' of Atkin and Swinnerton-Dyer

Developing the classical ideas of the theory of modular functions, Atkin and Swinnerton-Dyer in [40] place the theory of modular form for non-congruence subgroups in the modular group on fundamental issues. At first, the question is of examining algebraic functions with a given order of branching in accordance to Theorem 18. In the cases where the values d of index are not too large, algebraic functions can be constructed explicitly which leads to important formulas for the automorphic equation (24).

We now outline the method of Atkin and Swinnerton-Dyer of constructing so-called 'J-equations' in the context of the modular group. Part of the results can be carried over to a more general case, for example, to the triangular group from Theorem 11. Suppose first that the subgroup $\overset{\circ}{\Gamma}$ is cycloidal, i.e. $\overset{\circ}{h} = 1$. The classical modular invariant $j(z)$ has the Fourier expansion

$$j(z) = e^{-2\pi i z} + 744 + 196884\, e^{2\pi i z} + \cdots = \sum_{k=-1}^{\infty} A_k e^{2\pi i k z}, \tag{62}$$

where one can see that $j(z)$ has a simple pole at infinity relative to the local parameter $\tau = e^{2\pi i z}$. It is known that at the elliptic vertex of second order b_1 of the fundamental domain $j(b_1) = 1728$, at the elliptic point of third order b_2 $j(b_2) = 0$. We choose the Hauptfunktion for the subgroup $\overset{\circ}{\Gamma} \subset \Gamma_z$ to be normalized by the following conditions

$$\varsigma(s) = e^{-2\pi i z/d} + \underset{\operatorname{Im} z \to \infty}{O}(\exp -2\pi \operatorname{Im} z). \tag{63}$$

Recall that the question is of a cycloidal subgroup $\overset{\circ}{\Gamma}$ for which formula (48) takes the form

$$d = \mu_1.$$

For a cycloidal subgroup and invariants (62), (63) we have, evidently, only a polynomial relation instead of rational (24) which can be written in a more convenient form:

$$\begin{cases} j = F_2^3 \cdot E_2 \\ j - 1728 = F_1^2 \cdot E_1, \end{cases} \tag{64}$$

since we are interested in the order of branching at the points $j = 0$ and $j = 1728$. In formula (64) F_1, F_2, E_1, E_2 are polynomials in the Hauptfunktion $\varsigma(z)$. The degrees of $F_2^3 E_2$ and $F_1^2 E_1$ are equal to d. In addition, in accordance to the Riemann–Hurwitz formula (47) E_1 has the degree e_1, E_2 has the degree e_2, F_1 has the degree $e_1 + 2e_2 - 3$ and F_2 has the degree $e_1 + e_2 - 2$. Further concrete definition of the Equations (64) (J-equations) depends on the specific properties of the subgroup $\overset{\circ}{\Gamma}$.

We now consider an arbitrary subgroup $\overset{\circ}{\Gamma}$ of genus zero of the modular group with specification $(\overset{\circ}{g} = 0, d, \overset{\circ}{h}, e_1, e_2, \mu_1, \ldots, \mu_{\overset{\circ}{h}})$ (see Section 10). It is natural to suppose that $\overset{\circ}{h} > 1$. For writing J-equations in this case, Atkin and Swinnerton-Dyer introduce another specification of a given subgroup $\overset{\circ}{\Gamma}$. Namely, in addition to short specification $(\overset{\circ}{g} = 0, d, \overset{\circ}{h}, e_1, e_2)$ one takes a set of natural numbers α_j, ν_j, $1 \leq j \leq s$, $\nu_1 > \nu_2 > \cdots > \nu_s$ such that

$$\begin{cases} d - \mu_1 = \alpha_1\nu_1 + \alpha_2\nu_2 + \cdots + \alpha_s\nu_s \\ h - 1 = \alpha_1 + \alpha_2 + \cdots + \alpha_s \end{cases} \tag{65}$$

(see (47), (48)). 'J-equations' take the form

$$\begin{cases} j \prod_{i=1}^{s} A_i^{\nu_i} = F_2^3 E_2, \\ (j - 1728) \prod_{i=1}^{s} A_i^{\nu_i} = F_1^2 E_1, \end{cases} \tag{66}$$

where the polynomial A_i has the degree α_i in $\varsigma(z)$. In order for Theorem 18 to be valid, one must claim $A_1 \cdot A_2 \cdots A_s \cdot E_1 \cdot F_1 \cdot E_2 \cdot F_2$ not to have multiple roots. As in the case $\overset{\circ}{h} = 1$ further detailing (66) depends on characteristics of $\overset{\circ}{\Gamma}$. For small values of d these equations can be computed explicitly (see [1], [40]). Some examples for those $\overset{\circ}{\Gamma}$ appear in the following section.

14. Examples of Variable Changes with Controlled Monodromy Groups

All examples given here are well-known in another context and concern the subgroups of the modular group. The question is of the examples for computing automorphic equation (24) (or Equations (64), (66)) realizing a variable change in (5).

(a) We begin with the principal congruence subgroup $\Gamma(n)$. We naturally restrict ourselves to $n = 2, 3, 4, 5$ ($\overset{\circ}{g} = 0$). Automorphic equations which connect the invariants (Hauptfunktionen) for $PSL(2, \mathbb{Z})$ and $\Gamma(n)$ were calculated in [41]*. We give here these formulas for $n = 2, n = 5$ (see [42]).

Let the Hauptfunktion $J(z)$ for $\Gamma_{\mathbb{Z}}$ be normalized by conditions

$$J(\exp(2\pi i/3)) = 0, \quad J(i) = 1, \quad J(i\infty) = \infty,$$

in other words, $J(z) = j(z)/1728$. The Hauptfunktion $J_{\Gamma(2)}(z) = \lambda(z)$ is connected

* This book [41] was unavailable to me. I have taken part of its results from other references.

with $J(z)$ by the famous expression

$$J(z) = 4/27 \frac{\left(1 - \lambda(z) + \lambda^2(z)\right)^3}{\lambda^2(z)(1 - \lambda(z))^2}. \tag{67}$$

For the group $\Gamma(5)$ the Hauptfunktion $J_{\Gamma(5)} = \Lambda$ is connected with $J(z)$ by a rational function:

$$J(z) = -\frac{\left[u(\Lambda(z))\right]^8}{1728[w(\Lambda(z))]^5},$$

where $w(t) = t(t^{10} + 11t^5 - 1)$, $u(t) = t^{20} + 1 - 228(t^{15} - t^5) + 494t^{10}$.

(b) We now consider the case of the Hecke congruence subgroup. The complete list of values n with $\overset{\circ}{g} = 0$ is given in (30). As far as we are aware, the automorphic equations connecting the Hauptfunktionen $J_{\Gamma_{\mathbf{z}}}$ and $J_{\Gamma_0(n)}$ for all of n from the list (30) are not computed (or, at least, they are not published in the literature available to me), although the problem of computing these equations is classical.

There exists Newman's method (see [43], [44]) which enables one to construct analytical automorphic functions in the case of general group $\Gamma_0(n)$ in terms of Dedekind η-function (31).

THEOREM 19. *The function*

$$\prod_{d|N} \eta(dz)^{r(d)} \tag{68}$$

where the product is taken over all the divisors of the number N, $r(d) \in \mathbb{Z}$, is automorphic relative to $\Gamma_0(N)$ and is analytic, if the following conditions are fulfilled: (1) $\sum r(d) = 0$, (2) $\prod d^{r(d)}$ is a square (rational), (3) $\prod \eta(dz)^{r(d)}$ has the integer order at each cusp of the fundamental domain of the group $\Gamma_0(N)$.

The question arises from the choice of the numbers $r(d)$ in such a way that the function (68) becomes univalent. As Birch has noted (see [45]), in addition to the list (29) the function

$$J(z) = \left(\frac{\eta(N(z))}{\eta(z)}\right)^{24/(N-1)}$$

is the Hauptfunktion for $\Gamma_0(N)$ (see Section 8) for the values $N = 4, 9, 25$, or rather, for the values of

$$N = 2, 3, 4, 5, 7, 9, 13, 25. \tag{69}$$

Thanks to Theorem 19 the expression for $J_{\Gamma_0(n)}$ can be found in terms of η-functions for other values of N in the list (30). For example, for $N = 6$ and $N = 8$ we have

$$J_{\Gamma_0(6)}(z) = \frac{\eta(z)^5\eta(3z)}{\eta(2z)\eta(6z)^5}, \quad J_{\Gamma_0(8)}(z) = \frac{\eta(z)^4\eta(4z)^2}{\eta(8z)^4\eta(2z)^2}.$$

Now for computing automorphic equations

$$J_{\Gamma_{\mathbf{Z}}} = F(J_{\Gamma_0(n)}),$$

it suffices theoretically to construct the Fourier expansion for $J_{\Gamma_0(n)}(z)$ of the type (62), to calculate sufficiently many Fourier coefficients and to make use of the techniques of 'J-equations' (66). For this we must substitute these expansions in (66) and solve the algebraic system of equations in coefficients of J-equations. I think that this problem can be overcome, although it demands the help of a computer.

The computations are essentially simplified if we use some geometric arguments. Coming from (and only) the structure of the fundamental domains for $\Gamma_0(2)$, $\Gamma_0(3)$ Birch computed the corresponding automorphic equations (see [45]). We give here these formulas: for the group $\Gamma_0(2)$ the Hauptfunktion $J_{\Gamma_0(2)}(z) = t_2(z)$ is connected with the modular invariant $j(z)$ by the formula

$$j(z) = \frac{(t_2(z)+256)^3}{t_2^2(z)}.$$

The analogous formula for $\Gamma_0(3)$ has the form

$$j(z) = \frac{(t_3(z)+27)(t_3(z)+243)^3}{t_3^3(z)}.$$

(c) The examples of Atkin and Swinnerton-Dyer of solving 'J-equations'.

The following examples were found in [40]. The first example is related to cycloidal subgroups of $\Gamma_{\mathbf{Z}}$. We let a short specification be given in such a way

$$(\overset{\circ}{g}, d, \overset{\circ}{h}, e_1, e_2) = (0, 5, 1, 1, 2).$$

It is evident that this does not contradict the Riemann–Hurwitz formula (47). Then J-equations (64) take the form

$$j = \varsigma^3(\varsigma^2 + 5k\varsigma + 40k^2)$$
$$j - 1728 = (\varsigma^2 + 4k\varsigma + 24k^2)^2(\varsigma - 3k),$$

where $k^5 = 1$. The equations uniquely define 5 subgroups of the modular group which are all congruence subgroups and exhaust all subgroups in $\Gamma_{\mathbf{Z}}$ of index 5 (see (46)).

Another example given by the specification $\overset{\circ}{g} = 0$, $d = 9$, $\overset{\circ}{h} = 3$, $e_1 = 1$, $e_2 = 0$, $\mu_1 = 7$, $\nu_1 = 1$, $\alpha_1 = 2$ defines the set of 7 non-congruence subgroups of level 7 by the equations

$$j(\varsigma^2 + \tfrac{13}{4}k\varsigma + 8k^2) = (\varsigma^3 + 4k\varsigma^2 + 10k^2\varsigma + 6k^3)^3, \quad k^7 = 64,$$
$$(j - 1728)(\varsigma^2 + \tfrac{13}{4}k\varsigma + 8k^2) = \varsigma(\varsigma^4 + 6k\varsigma^3 + 21k^2\varsigma^2 + 35k^3\varsigma + 63k^4/2)^2.$$

(d) Classical examples connected with normal extensions over the fields of automorphic functions.

We now recall that there exists the list (53) of all normal subgroups of the modular group with $\overset{\circ}{g} = 0$. We have considered the automorphic equations for the subgroups $\Gamma(2)$–$\Gamma(5)$. Automorphic equations for the groups $\Gamma^2_{\mathbf{Z}}$, $\Gamma^3_{\mathbf{Z}}$ have the form

$$j(z) = J^2_{\Gamma^2_{\mathbf{Z}}}(z) + 1728, \quad j(z) = J^3_{\Gamma^3_{\mathbf{Z}}}(z).$$

Another interpretation of automorphic equations for normal subgroups of $\Gamma_{\mathbf{Z}}$ is connected with Galois extensions over the field of modular functions. In particular, in the examples just considered one can say on the existence of normal extensions $K(j^{1/3})$, $K(j - 1728)^{1/2}$ over the field of modular functions with Galois groups isomorphic to factor groups $\Gamma_{\mathbf{Z}}/\Gamma^3_{\mathbf{Z}}$, $\Gamma_{\mathbf{Z}}/\Gamma^2_{\mathbf{Z}}$.

Our last example is connected with the normal extension over the field of automorphic functions for the group $\Gamma(2)$. Namely, this extension is defined by the root

$$\lambda^{1/n}(z),$$

where $\lambda(z)$ is the Hauptfunktion for $\Gamma(2)$ (see (67)). There exists a normal subgroup $\overset{\circ}{\Gamma}_n$ in $\Gamma(2)$ of genus zero such that its Hauptfunktion $J_{\overset{\circ}{\Gamma}_n}(z)$ is connected with $\lambda(z)$ by the relation

$$\lambda(z) = J^n_{\overset{\circ}{\Gamma}_n}(z). \tag{70}$$

As it follows from [31], [32], [37], [38] this group $\overset{\circ}{\Gamma}_n$ can be defined by formula (59) or can be given by the system of generators (57). As previously noted, only for $n = 1, 2, 4, 8$ $\overset{\circ}{\Gamma}_n$ is the congruence subgroup in $\Gamma_{\mathbf{Z}}$. Combining (67) and (70), we obtain the automorphic equation

$$J(z) = J_{\Gamma_{\mathbf{Z}}}(z) = \frac{4}{27}\frac{\left(1 - J^n_{\overset{\circ}{\Gamma}_n}(z) + J^{2n}_{\overset{\circ}{\Gamma}_n}(z)\right)^3}{J^{2n}_{\overset{\circ}{\Gamma}_n}(z)\left(1 - J^n_{\overset{\circ}{\Gamma}_n}(z)\right)^2}.$$

References

1. Klein F. and Fricke R., Vorlesungen über die Theorie der elliptischen Modulfunktionen I, II, Teubner, Leipzig (1890, 1892).
2. Klein F. and Fricke R., Vorlesungen über die Theorie der automorphen Funktionen I, II, Teubner, Leipzig (1897, 1912).
3. Lappo-Danilevskiĭ I. A., Mémoires sur la théorie des systèmes des equations differentialles linéaires. Chelsea Publ. Comp. (1953).
4. Venkov A. B., Constructing the Hauptfunktion, solving the Schwarz and Fuchs equations for a surface of genus zero by the method of spectral theory of automorphic functions. In the book: Differential geometric, Lie groups and mechanics, *Zap. Nauchn. Semin. Leningrad. Otdel. Mat. Inst. Steklov* (LOMI), **133** (1984), 51–62 (Russian).
5. Venkov A. B., On explicit formulas for accessory coefficients in the Schwarz equation. *Funktional Analysis and Its Appl.*, **17** (1983) N 3, 1–8 (Russian).
6. Venkov A. B., Spectral theory of automorphic functions and its applications.- In: *Proc. of ICM in Warsaw, Polish Scientific Publ.* (1984), v. **2**, 909–919.

7. Lehner J., Note on the Schwarz triangle functions. *Pacific Journ. Math.* (1954), **4**, 243–249.
8. Beardon A. F., The geometry of discrete groups. *Graduate Texts in Math.*, **91**, Springer-Verlag (1983).
9. Takeuchi K., Arithmetic triangle groups, *J. Math. Soc., Japan* (1977), v. **29**, 91–106.
10. Helling H., On the commensurability class of the rational modular group. *J. London Math. Soc.* (1970), v. **2**, 67–72.
11. Kluit P. G., Hecke operators on $\Gamma^*(N)$ and their traces, Dissertation. Vrije Universiteit te Amsterdam (1979).
12. Swinnerton-Dyer H. P. F., Arithmetic groups. In the book: Discrete groups and automorphic functions, edited by W. Harvey, Academic Press (1977), 377–401.
13. Raleigh J., On the Fourier coefficients of triangle functions, *Acta Arithm.* (1965), v. **8**, 107–111.
14. Apostol T. M., Modular functions and Dirichlet series in number theory, *Graduate Texts in Math.*, v. **41**, Springer–Verlag (1976).
15. Kurosh A. G., Group theory. M. (1953).
16. Dey I. M. S., Schreier systems in free product. *Proc. Glasgow Math. Assoc.* (1965), v. **7**, 6–79.
17. Newman M., Asymptotic formulas related to free products of cyclic groups. *Maht. Comput.* (1976), v. **30**, N 136, 838–846.
18. Stothers W. W., Free subgroups of the free product of cyclic groups. *Math. Comput.* (1978), v. **32**, N 144, 1274–1280.
19. Millington M. H., Subgroups of the classical modular group. *J. London Math. Soc.* (1969), v. **1**, 351–357.
20. Singerman D., Subgroups of Fuchsian groups and finite permutation groups. *Bull. London Math. Soc.* (1970), v. **2**, 319–323.
21. Kulkarni R.S., A new proof and an extension of a theorem of Millington on the modular group, Preprint MPI/SFB 84-49, Max-Planck Institut für Math., Bonn (1984).
22. Kulkarni R. S., An extension of a theorem of Kurosh and applications to Fuchsian groups. *Michigan Math. J.* (1983), v. **30**, 259–272.
23. Jones G. A., Singerman D., Theory of maps on orientable surfaces. *Proc. London Math. Soc.* (1978), v. **37**, 273–307.
24. Petersson H., Konstruktionsprinzipen für Untergruppen der Modul-gruppe mit einer oder zwei Spitzenklassen. *J. Reine Angew. Math.* (1974), v. **268/9**, 94–109.
25. Stothers W. W., Subgroups of the modular groups. *Proc. Camb. Phil. Soc.* (1974), v. **75**, 139–153.
26. Stothers W. W., On a result of Petersson concerning the modular group. *Proc. Roy. Soc. Edinburgh* (1981), v. **87 A**, 263–270.
27. Stothers W. W., Impossible specifications for the modular group. *Manuscripta Math.* (1974), v. **13**, 1–15.
28. Newman M., Classification of normal subgroups of the modular group. *Trans. Amer. Math. Soc.* (1967), v. **126**, 267–277.
29. Newman M., A complete description of the normal subgroups of genus one of the modular group. *Amer. J. Math.* (1964), v. **86**, 17–24.
30. McQuillan D. L., Classification of normal congruence subgroups of the modular group. *Amer. J. Math.* (1965), v. **87**, 285–296.
31. Fricke R., Ueber die Substitutionsgruppen, welche zu den aus dem Legendreschen Integralmodul $k^2(w)$ gezogenen Wurzeln gehören. *Math. Ann.* (1887), v. **28**, 99–118.
32. Pick G., Ueber gewisse ganzzahlige lineare Substitutionen, welche sich nicht durch algebraische Congruenzen erklären lassen. *Math. Ann.* (1887), v. **28**, 119–124.
33. Newman M., Normal subgroups of the modular group which are not congruence subgroups. *Proc. Amer. Math. Soc.* (1965), v. **16**, 831–832.
34. Wohlfahrt K., An extension of F. Klein's level concept. *Illinois J. Math.* (1964), v. **8**, 529–535.
35. Knopp M. I., Newman M., Congruence subgroups of positive genus of the modular group. *Illinois J. Math.* (1965), v. **9**, 577–583.
36. McQuillan D. L., On the genus of fields of elliptic modular functions. *Illinois J. Math.* (1966), v. **10**, 479–487.
37. Sansone G., Problemi insoluti nella teoria delle sostituzioni lineari. Conv. Inter. di Teoria dei Gruppi Finiti (Firenze 1960) Rome (1960).
38. Newman M., On a problem of G. Sansone. *Ann. di Mat. Pura Appl.* (1964), v. **65**, 27–33.
39. Jones G. A., Triangular maps and non-congruence subgroups of the modular group. *Bull. London Math. Soc.* (1979), v. **11**, 117–123.

40. Atkin A. O. L., Swinnerton-Dyer H. P. P., Modular forms on non-congruence subgroups. *Proc. Sympos. Pure Math. AMS* (1971), v. **19**, 1–26.
41. Fricke R., Lehrbuch der Algebra, vol. **2**, Braunschweig (1926).
42. Bateman H. and Erdelyi A., Higher transcendental functions, v. **3**, McGraw-Hill, New York (1955).
43. Newman M., Construction and application of a class of modular functions, I, *Proc. London Math. Soc.* (1957), v. **7**, 334-350.
44. Newman M., Construction and application of a class of modular functions, II, *Proc. London Math. Soc.* (1959), v. **9**, 373–387.
45. Birch B., Some calculations of modular relations, Lect. Notes in Math., Springer-Verlag (1973), v. **320**, 175–186.

Author Index

Subject Index

9 780792 304876